THE BRAIN

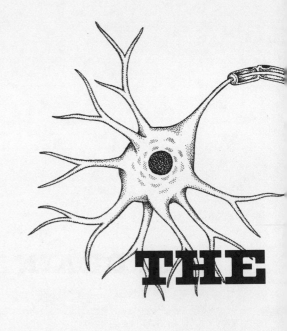

THE

Rob DeSalle &
Ian Tattersall

Illustrated by Patricia J. Wynne

Yale UNIVERSITY PRESS

NEW HAVEN & LONDON

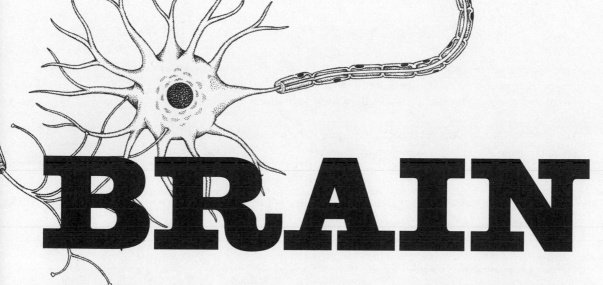

BRAIN

Big Bangs, Behaviors, and Beliefs

Published with assistance from the Louis Stern Memorial Fund.

Yale University Press books may be purchased in quantity for educational, business, or promotional use. For information, please e-mail sales.press@yale.edu (US office) or sales@yaleup.co.uk (UK office).

Designed by Lindsey Voskowsky.
Set in Joanna MT type by Westchester Book Group.
Printed in the United States of America.

Library of Congress Cataloging-in-Publication Data

DeSalle, Rob.
   The brain : big bangs, behaviors, and beliefs / Rob DeSalle and Ian Tattersall ;
illustrated by Patricia J. Wynne.
      p.   cm.
   Includes bibliographical references and index.
   ISBN 978-0-300-17522-6 (clothbound : alk. paper)   1. Cognition.
2. Neurophysiology.   3. Brain—Evolution.   I. Tattersall, Ian.   II. Title.
   BF311.D466 2012
   612.8'2—dc23   2011044329

A catalogue record for this book is available from the British Library.

This paper meets the requirements of ANSI/NISO Z39.48–1992 (Permanence of Paper).

10 9 8 7 6 5 4 3 2 1

For Erin, Jeanne, and Maceo

CONTENTS

PREFACE

You don't have to peer expensively out into the farthest reaches of the cosmos to encounter the deepest mystery in the universe. Look no further than right between your own two ears. The human mind is nowhere challenged as much as by the endeavor to understand the workings of the brain that produces it. Yet despite huge practical difficulties, the effort to comprehend how brain produces mind is perhaps more intrinsically fascinating than any other quest in science. The many other organs that make up our bodies can be pretty well understood as machines, devoted to fairly clearly defined, if complex, functions. The special challenge posed by the brain-machine is that within it there also resides a ghost: our emergent consciousness, the properties of which far exceed the sum of the internal connections and electrochemical discharges that give rise to it.

The same is also true, at least to some extent, for the "brain" of every other creature that possesses one: after all, each has a consciousness of some kind, if only the rudimentary ability to distinguish between "self" and "other." But there is nonetheless something very special about the human brain. As far as we can tell, it is alone in its ability not only to respond to the stimuli that come in from the world around it but also to literally remake that world in imagination. Nobody knows exactly how it accomplishes this trick, but a huge amount is already understood, not only about its structure and functioning but about the wider zoological and evolutionary contexts in which the amazing modern human brain and its unique properties need to be comprehended. These are the subjects of this book and of the American Museum of Natural History exhibition (*Brain: The Inside Story*) that inspired it.

In their structure our brains bear the marks of a long and tortuous ancestry. These odd machines are something that no human engineer would ever have designed; and if we understood nothing of their complex and accretionary half-billion-year history we would be at a loss to explain their apparent untidiness. We need, then, to analyze them in evolutionary terms. But if an evolutionary approach to understanding the human brain is to make even a little sense, we also need to know where we fit into the larger scheme of life. This is because the living world around us, together with the ways in which it is organized, offers a host of clues to how we achieved our unusual intellectual eminence. Yet even the combination of knowing both our place in nature and how we got there is not quite enough. We need also to understand the underlying evolutionary mechanisms.

Most of us were taught in school—if indeed the subject was even mentioned—that evolution basically involves steady improvement over the eons. But it has become widely appreciated, within our own academic lifetimes, that there is a lot more to evolution than this. And once we realize that evolution is not always the fine-tuning mechanism many of us thought it was, we can begin to appreciate more fully the facts of our biological history as human beings.

These basic facts are provided by our fossil record, the preserved physical remains of our predecessors. Among other things, fossils show how the human brain has dramatically expanded in size over the past two million years or so. And, almost uniquely in the human case, there is more. Because, for the past two and a half million years or more that have elapsed since the invention of stone tools, we also have the archaeological record, the direct register of the behaviors of our hominid predecessors. By comparing our physical and behavioral records we can better understand past patterns of innovation in human evolution, and we can begin to reconstruct the framework within which the extraordinary human spirit emerged.

Still, we cannot do this without first understanding the other kinds of brains that exist on earth—and indeed, appreciating how it is that some living things get by perfectly well without any brains at all. And we also need to know how our brain works today. Exactly how our unique human consciousness is generated remains for the time being a mystery. But a huge amount is already understood about brain processes at the molecular, anatomical, and functional levels; and new techniques of imaging have given us unparalleled insights into our "brains at work." These technologies allow us to see in real time what actually goes on in human brains when they are subjected to specific stimuli or are set to performing particular tasks. We have also developed an advanced appreciation of the "layered" structure of the brain and of how this relates to the increasingly complex functions that vertebrate brains have acquired over the eons.

In this book we will look at all of the factors that led to our ability to think about thinking, and we will see just how unique our modern way of processing mental information is. Thanks to some remarkable new developments, we will gain insight into the true complexity (and murkiness) of the decision-making processes that go on in our brains, and we will look at how the untidiness (and creativity) of our mental processes results from the long and eventful evolutionary history of the brain. We will see how the qualities of our minds that we prize the most are "emergent," rather than the fine-tuned products of inexorable natural selection.

We will also look toward the future of our brains. We are the recently arrived and sole hominid inheritors of the earth. Yet in our short tenure we have made a more profound impact on the planet than any other species has ever contrived to do. Most of the negative effects of human activity are by-products: the unintended consequences of a reckless attitude. Can we expect evolution to burnish us into better stewards of our environment? Can we hope for genetic engineering to make us more responsible, or more efficient, or simply better? Or are we going to have to learn to live with ourselves as we are? We hope to show that we are capable of understanding these important questions as long as we keep them in evolutionary context; and we may, indeed, even be capable of dealing with the consequences of our activities as long as we recognize and compensate for the untidiness of our thought processes.

ACKNOWLEDGMENTS

We thank Lauri Halderman and her co-writers, Martin Schwabacher, Margaret Dornfeld, and Sasha Nemecek, for their astute and careful crafting of the label copy for the exhibition *Brain: The Inside Story*. We quote their writing several times in this book, and these excerpts show the clarity with which they can make difficult subjects understandable. We also thank David Harvey for his splendid vision in designing the exhibition, and for so successfully shepherding it to completion. Among the many other contributors to the exhibit, Helene Alonso deserves special recognition for her creative interactives, which we mention several times. On the scientific side, our gratitude goes to Rob's two scientific collaborators, Joy Hirsch and Maggie Zellner, who co-curated the exhibition. Without their input, our understanding of much of the neuroscience we discuss in this book would have been much less than desirable. Our thanks also go both to Vivian Schwartz, museum volunteer and explainer in the *Brain* exhibition, who read the manuscript and gave us great initial comments, and to Rob's "Bring Your Daughters to Work" intern Ashley Emmell, whose expert organizational abilities helped us immensely with the drawings in this book. And of course our appreciation goes to the thousands of museum patrons who viewed the exhibition during its run at the American Museum. It is their enthusiasm and interest in the exhibition that really fueled the writing of this book.

At Yale University Press our immense appreciation goes to our editor, Jean Thomson Black, whose enthusiasm got the project under way and whose editorial suggestions vastly improved the final product, to Laura Jones Dooley for her immaculate copyediting, to Jesse Hernandez for other refinements, and to Sara Hoover for her unfailing helpfulness and efficiency in dealing with the details. We thank Don McGranaghan, Diana Salles, Jennifer Steffey, Willard Whitson, and Gisselle Garcia for their help with the illustrations in chapter 10. And finally, we would like to acknowledge what a pleasure it has been to work once again with the principal illustrator of this book, Patricia J. Wynne. Her illustrations are charismatic, and her input invariably goes far beyond the visual.

THE BRAIN

1 THE NATURE OF SCIENCE

Our Brains at Work

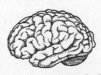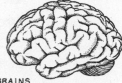

HOMINID BRAINS

What does it mean to think about thinking? Only members of our species can ask this question. No other organism on this planet has the physical or neural makeup to come even close to posing it. And the answer involves a sometimes convoluted intellectual journey, albeit one with its destination steadily in the human brain. Understanding just how our unique modern human style of cognition was acquired necessitates looking at evidence of many kinds, much of it gathered from the study of a huge diversity of living organisms, but always in an evolutionary context. The evolutionary approach culminates in the viscerally most dramatic story of all—how, after several million years of jostling for ecological space with a host of other hominid species, a single lineage of our zoological family managed to eliminate all the competition, until there was only one actor left standing on the human evolutionary stage: us.

Our brains govern virtually every action we take. Behaviorally, it is our brains that make each of us what we are as unique individuals. And collectively, it is the extraordinary and totally unprecedented human brain that enables our species to be the psychologically complex, highly distinctive, and occasionally bizarre entity that it is. This book is about this delicate, elaborately configured, and yet highly adaptable mass that resides within our oddly large cranial vaults and that allows each of us to experience the world in the unique human fashion. It is about the simple origins and long evolutionary history of the brain: how it is structured, how it functions and changes through life, and how we have come to know what we do—so much, and yet so little—about this most mysterious of organs.

An Inelegantly Contrived but Rather Effective Evolutionary Mess

Much as we might wish to vaunt our modern human brains that have allowed our species to achieve so much, the New York University cognitive psychologist Gary Marcus has colorfully likened them to "kluges": the inelegant but effective solutions to problems that are typically produced by garage tinkerers. This is because, as we'll see, on examination they turn out to be a basically jury-rigged solution to a long succession of evolutionary problems. The human brain is hardly alone in this, for the "messy" organ that resides in our heads is in fact just one example of the kind of untidy history that most of the structures of living organisms accrue; but it may be the most wonderful example of all.

If one were to design a brain from scratch, it certainly wouldn't look or work at all like our brains. It might look a lot like a modern-day computer, or something more elegant yet, but it most assuredly would not be the convoluted Rube Goldberg apparatus we have inside our heads. As a simple example of how inefficiently our brains are designed, we need only look at how we store memories. The easiest way to store something is to give it an address, place it where it belongs in storage according to this address, and keep track of all of the addresses. Librarians do this, museum scientists do it, and a lot of modern human daily life consists of giving things addresses so we can store them logically for easy retrieval, whether on notecards, sticky notes, or Excel spreadsheets or in computer desktop folders. Indeed, our computers are probably the epitome of how the human brain *ought* to store things. But even though human brains know enough to design computers as efficient information storage and retrieval systems—and indeed they have worked out very effective and nearly perfect methods in this domain—the memories that reside inside our brains themselves are not stored efficiently at all. This is why we forget so many things and why our memories are so frequently inaccurate. If recent estimates of the unreliability of eyewitness testimony at criminal trials are accurate, then our brains are pretty awful at storage and retrieval.

Memory, what's more, is only one example of our mental inefficiencies: many other aspects of our brains are even less logically constructed. Why? The reason is simple, and it is a historical one. We humans are only a single species in a great tree of life that has diverged and branched from a single common ancestor that lived more than three and a half billion years ago. As a result of this three-and-a-half-billion-year experiment, our genomes—and the genomes of every organism that has ever lived on this planet—carry data about our past, our present, and even our future potential. As we'll see, evolution is not the process of optimization it is often thought to be but is instead often hostage to history and chance. And this is why our brains are such a mess.

Thinking about Thinking

Let's start this book by posing the question of how we go about trying to comprehend the human brain and its properties, because it is crucial to understanding everything else that follows. All fields that contribute to our knowledge of the brain as an evolutionary and functional entity are scientific disciplines. And many of the products of the brain—for example, fiction, poetry, and spiritual ideas—are hardly susceptible to scientific analysis, because science deals solely with observations we can make about the material world. But to the extent that the human brain itself is a functioning material object, upon which repeatable observations can be made, it is indeed an appropriate subject—perhaps the *ultimate* subject—for the scientific way of comprehending the world that the unique properties of the human brain make possible.

Scientific understanding differs from every other kind of knowing, in being explicitly *provisional*. Perhaps one of the greatest myths of the modern age is that science is an authoritarian system, piling up mountains of immutable objective facts. The reality is quite the reverse. Because although it depends on human creativity and intuition as much as any other intellectual endeavor, science is founded on, and is most clearly distinguished by, the doubt and questioning it entails. Indeed, the oft-heard phrase "scientifically proven" is one of the most misleading clichés around. Science is not actually about proving anything. Instead, scientists gradually edge in on a more precise and accurate description of nature by proposing ideas about it (which is where the creativity comes in) and then discarding those that do not stand up to scrutiny (which is where doubt enters the picture). For scientists to do this, of course, those ideas have to be explicitly testable, so to be scientific an idea has to be formulated with its potential falsifiability in mind. If an idea is not somehow testable in the observable world, then it is not a strictly scientific one (though the larger fabric of science may also weave in notions that may not be directly testable in practical terms, at least with the equipment we have currently at hand).

Some scientific ideas have withstood examination by generations of scientists for decades or even centuries; others may be falsified tomorrow. If they are disproved, they will then have to be rejected, no matter how clever they are or how attached their authors may be to them. This process of proposing and testing hypotheses makes science the ultimate collective activity: it is a lot harder to imagine a world with just one scientist than to imagine one without any scientists at all. In the end, science is a self-correcting system that depends on its own internal system of vigilance, a system that is maintained by an essential plurality of ideas. This is actually one of the things that makes being a scientist so cool: unlike an engineer or a physician, whose decisions have to be right or the consequences may be disastrous, a scientist may be wrong and still feel that he or she is contributing valuably to an ongoing process—a process that is far bigger than any one individual. The corollaries of never knowing for sure that you are right and having to admit it immediately when you're proven wrong are small prices to pay for this kind of satisfaction!

Scientific knowledge is thus a constantly moving target. It is a process, not a durable product. For the product itself is always changing—which is, after all, what the huge and exponentially growing scientific literature is all about. What's more, contrary to common belief, even the process itself is not unified by the "scientific method" that we constantly hear about. The great biologist Peter Medawar, perhaps the finest scientific writer of all time, made our favorite comment on this subject in 1969: "Ask a scientist what he conceives the scientific method to be and he will adopt an expression that is at once solemn and shifty-eyed: solemn, because he feels he ought to declare an opinion; shifty-eyed, because he is wondering how to conceal the fact that he has no opinion to declare." The reality is that there is not one scientific method but many, each specifically designed for its suitability to the phenomenon being investigated. Rather than a rigid formula that dictates how we should proceed, science is simply an approach to knowledge.

Still, despite the amorphousness of the larger scientific process, it is entirely fair to conclude that, as a result of its workings, the picture of the universe and its contents that we have today is more accurate than the one we had yesterday—even if it is almost certainly less accurate than the one we will have tomorrow. And the sum total of scientific knowledge at any one point in time definitely gives us a platform that is firm enough to allow us to build on it, constantly and confidently, toward tomorrow's understanding. Indeed, some scientific ideas and observations of nature have proven so durable and resistant to falsification that we may regard them as the scientific equivalent of fact—although testing will always continue.

Shared, Derived. Derived, Shared. That Is All Ye Know on Earth and All Ye Need to Know

One of those robust hypotheses about the world is the notion of evolution. A century and a half after it was proposed, evolution is still the only scientific hypothesis we have that actually *predicts* the organization we observe in the living world—namely, that organisms visibly fall into a system of sets-within-sets that can be represented by a branching diagram (a "tree") with repeating forks. This pattern has been corroborated time and again, as our knowledge of the biosphere has grown by leaps and bounds.

Today we trace the beginnings of modern evolutionary biology to the year 1858, when the evolutionary views of the British naturalists Charles Darwin and Alfred Russel Wallace were presented to a meeting of the Linnean Society in London. This historic event was hardly appreciated at the time for the turning point it was, but it was followed the next year by the publication of Darwin's magisterial book *On the Origin of Species by Natural Selection*, which really did take the world by storm. Darwin's own thumb-nail characterization of evolution was "descent with modification," a very neat way of expressing that all living organisms, no matter how disparate, are connected by ancestry. It is this which gives rise to the clear pattern of nature we've just noted.

To start at the base of the tree that describes this pattern, a single ancient common ancestor gave rise to the three major branches, or domains, of life: Bacteria, Archaea, and Eukarya. Each of these great groups then diversified repeatedly in the same way to produce the huge profusion of subgroups we see today. Eukarya (eukaryotes), for example, consists of numerous clusters of sometimes very different-looking organisms that are nonetheless united by possessing complex cells with membrane-bound nuclei. The nucleus was, of course, a feature that was present in the common ancestor of the group as a whole. Eukaryotes include, among certain other organisms, all of the animals and plants, ranging from simple unicellular creatures through mosses, green plants, fungi, sponges, flatworms, starfish, sharks, frogs, snakes, and crocodiles to platypuses and humans. Each of these and other large eukaryote subgroups is in turn subdivided, again and again, until we end up with literally millions of species.

We recognize the sequence of evolutionary splits that gave rise to this almost unimaginable diversity by looking for "derived" features (such as that membrane-bound nucleus in the case of the eukaryotes) whose pos-session binds the various groups and subgroups together. Any feature that is shared by all members of a particular group, to the exclusion of all others, will in principle be there by virtue of inheritance from the com-mon ancestor of that group—and will thus testify to the group's genea-logical unity. Besides the cell nucleus, another classic derived feature is provided by the feathers of birds—the discovery of which in certain amaz-ingly preserved dinosaur fossils has also recently allowed us to corroborate the early notion that these two groups are closely related. Characteristics like this permit us to recognize the boundaries of each group by exclud-ing from it forms that do not possess them or that at least are not descended from ancestors that possessed it; they also serve to admit new potential members to the group as they are discovered.

In contrast, the many "primitive" features of the common ancestor that were not unique to it will undoubtedly be found more widely than just in the descendant group. They thus cannot be used to determine membership in that group—even if all group members possess it. So although a cursory inspection of their body form would almost certainly suggest to you that a salmon and a lungfish were more closely related to each other than either is to a cow, it turns out on examination that the lungfish and the cow belong to a single group that shares a much more recent common ancestor than either has with the salmon. The lungfish and salmon look superficially similar, because they still share the watery habitat that was occupied by the even more remote ancestor of all three. The cow has become so distinctive by virtue of adaptation to an entirely new environment, while the salmon and the lungfish, sitting on independent branches of the tree, remained relatively unchanged. Additional complicating factors in figuring out the geometry of the great Tree of Life are found where organisms have changed so much over long intervals of time that there are few recognizable features left to link them together or, in cases of "convergence," where remotely related creatures have evolved remarkably similar adaptations.

There are traits, then, in organisms that can trick us into thinking they are the same thing. These are what Darwin called "analogies." Today we call these traits convergences or, in the jargon of the tree-builder, "homoplasies." Such traits arise independently in unrelated lineages. As an example, think of wings. There are many kinds of wings among animals. In fact, even some plant gametes can glide, float in the air, and move around by flying about. But let's look at three kinds of animal wings: those of birds, bats, and flies.

It is very easy to see that the fly wing is not "the same" as the bat or bird wing. It has no bones to support the flap of tissue that is used as the flying device, and its musculature is not derived from the same precursor cells as those of vertebrates (the bird and the bat). So we can easily say that fly wings (and indeed all insect wings) are not the same as vertebrate wings.

Analogy of animal wings. A bat wing, a bird wing, and an insect wing. Although these structures all have the same function (transportation of the owner through the air), they are not derived from the same ancestral structure. Hence, they are convergences, or homoplasies, with respect to one another.

The bat and bird wings are a little harder to interpret, but it's still pretty easy to see that they aren't equivalent except in function because the arrangements of the bones of the wing and their origins from precursor cells are completely different. So, from looking closely at the makeup of traits, we can get an idea of what is really "the same" and what is not.

The easiest way to go about determining the nature of the "sameness" involved is to look at the relationships of the species involved and to interpret the evolution of the trait we call "wings" in the context of those relationships. It is clear that, among the flies, birds, and bats, the birds and bats are more closely related, leaving the flies as "odd men out." When we look at just these three kinds of organism, we see that having a wing is a good defining characteristic. But there are actually millions of other species out there, most of them insects, that we should add to this smaller tree. And if we do so, we notice that just having wings no longer defines our fly-bird-bat group. This is because there are a lot of creatures more closely related to vertebrates that do not have wings (such as sea urchins and starfish). Consequently, the best explanation for wings in insects is that they arose independently in the common ancestor of the winged insects. Next, we can look at the pattern for bird and bat wings. Birds are a subgroup of dinosaurs that happen to have wings. Bats, by contrast, are mammals, and the common ancestor of dinosaurs and mammals most certainly did not have wings. We need only to look at the marsupials and the primitive mammals to realize that these forms don't have wings, and so the chain of common ancestry is broken, meaning that wings in birds arose independently of what we describe as "wings" in bats.

Could the Tree of Life Be a Periodic Table for Evolution?

In all of these cases, a powerful tool for resolving uncertainty is provided by the relatively recently developed science of molecular systematics ("molecular" refers to the hereditary molecule DNA that is the focus of this branch of science—we will hear a lot more about DNA later—while "systematics" is the name given to the study of the diversity of organisms and of the relationships among them). Serendipitously, and rather unexpectedly, it has turned out that certain parts of the long DNA strand are incredibly conservative and resistant to change over vast periods of time, so that distantly related organisms that have changed hugely in their physical aspect since they last had a common ancestor may actually share remarkably similar genes that can be directly compared. Where convergence has occurred, on the other hand, the structure of the genes responsible for the superficial similarities involved is likely to betray the underlying history involved.

The smallest individual unit into which nature is divided is the individual organism, which lives and dies independently of everything else that is out there. But from the point of view of someone trying to reconstruct evolutionary histories, the most important unit is the species. Scientists have never agreed on a definition of what species are (there are currently almost thirty definitions to choose from), but the most useful way to look at them is as the largest freely interbreeding populations of individuals. We members of the human species *Homo sapiens*, for example, don't exchange our genes with any other organisms (although in the past there might have been some biologically insignificant interchange with close relatives that are now extinct). Today we are an independent entity, a distinct actor in the evolutionary play. We might go extinct in competition with other organisms, but we will never be engulfed by one of them. Genera, in turn, are groupings composed of closely related species that have descended from a single ancestral species by what is known as "speciation." Over the eons, the splitting of lineages has been far from a rare event: there is, for example, a surprisingly large number of now-extinct species of our genus *Homo* known in the fossil record of the past two million years.

Fossils are the remains of ancient animals (in most cases only their tough teeth, carapaces, and bones) that have been preserved in sedimentary rocks laid down in the sea or, on land, mostly at the margins of rivers and lakes. These rocks accumulate as particles are eroded from preexisting rocks, to be transported by water or wind to new resting places on the landscape where they may cover and protect the weathering remains of dead organisms. In this way, a record of life is created as fossil-containing sediments are laid down, the younger on top of the older. Onshore fossils usually provide a highly sporadic record of past life, but marine sequences are often quite continuous over long periods of time. The end result is an admittedly incomplete record of earlier life; but nonetheless the fossil record is huge, and it is the only direct evidence we have of the history of life. We can work out the genealogical relationships of living organisms from their physical and molecular similarities, and from their molecular characteristics we can even make educated guesses about how long ago different organisms shared a common ancestor. But only the fossil record gives us a material record of the actual actors and events in the long evolutionary play that has unfolded over the past 3.5 billion years. Together, the fossil and comparative records give us an extremely powerful tool not only for corroborating the notion of evolution itself but also for re-creating the dramatic evolutionary story that has given rise to the marvelous and varied living world we see around us today.

Using the physical and molecular tools at our disposal, we can construct a Tree of Life that expresses the genealogical relationships among the millions of species now existing in the world. Starting at the tips of the tree, we can assemble species into the genera to which they belong. We can then combine the genera into increasingly comprehensive groups, until all living things are included in a single grand scheme. Traditionally, this scheme was represented by a vertical forking diagram that looks rather like an actual tree, with a basal trunk yielding to branches that in turn yield to branchlets and, finally, to twigs. But when as many groups are involved as there happen to be in the entire Tree of Life, it's often more convenient to represent them in a circular diagram, with younger divergences occurring farther from the center.

Our current scientific way of classifying living organisms was invented by the Swedish naturalist Carolus Linnaeus (Karl Linné) a century before Darwin and Wallace came up with the concept of evolution. Linnaeus was able, purely on the basis of physical resemblances, to devise a system whereby every species was grouped with others into a genus, genera were grouped into Orders, Orders into Classes, and Classes into Kingdoms. In contrast to the "military" style of hierarchy, where an individual can belong to only one rank, in the kind of "inclusive" hierarchy Linnaeus established, each species belongs to all of the ranks above it. Thus, as classified by Linnaeus, *Homo sapiens* resides not only in the genus *Homo* but also in the Order Primates, in the Class Mammalia, and in the Kingdom Animalia. Many other such "taxonomic ranks" have been added since Linnaeus's day to help us express the bewildering diversity of life on earth; gratifyingly, however, this preevolutionary basic arrangement still fits perfectly with the pattern of resemblance that we would expect to be produced by the evolutionary process, in which an ancient ancestral species gives rise to several descendant species, each of which diversifies in the same way, in a never-ending chain.

We suggest that the simple process of categorizing and naming objects around us that are relevant to our existence is in itself science; and so, in some sense, scientists must have existed since the beginning of language. The logical thought processes involved in determining that an object in nature has an essence, and hence can be called something, is an important scientific process that expresses itself in the guise of taxonomy. The journalist Carol Yoon has recently revived the term "umwelt" to describe the very basic need that humans have to name things. The umwelt refers to an understanding that we humans are part of the natural world; and this realization is so basic to our existence that in order to place ourselves in the natural world, we *need* to have names for things. Otherwise, we couldn't make sense of anything. The umwelt is like a nagging fly, buzzing about our heads, constantly reminding us that we are part of a natural system that *requires* explanation. The only way we can deal with it is to name things and to use those names to help provide understanding.

The science behind naming things is quite simple, and its philosophical underpinnings are at the heart of how we have always reasoned and thought about the natural world. Let's say you are a human being living in Africa twenty thousand years ago. You see an organism. The appearance of this organism is transmitted to your brain and processed to give you an "image" of it. Using what you know or remember from your experiences, you then dissect its components in your mind. You assess whether some of the parts of the organism are unique or shared with other organisms. In your mind, you ask whether various aspects of the organism are important enough for you to give them names. If some of the aspects are found in a lot of other organisms, they tell you that this one belongs to a bigger group of things. So if it has four legs, it goes with the other tetrapod animals. But while in some respects you might say to yourself, "I think it is like something I have seen before," others among its features might elicit the response, "I think it is not like anything I have seen before." In the light of these alternative hypotheses you make more observations, and eventually you convince yourself that it is something worth naming—basically as a result of testing the hypothesis that you have found a new organism. It turns out that the name you give the organism is yet another hypothesis and is itself subject to further testing. So even as an early human you are doing science: making hypotheses about nature that are meaningful and testable.

Of course, the amount of room in the world is finite, so in tandem with the process of diversification there has also been a vigorous pruning of branches, as huge numbers of species have succumbed to extinction on the ecological stage. But all in all, it is amazing how the basic sequence of historical events has managed to find itself preserved in the diversity of the living world. Around us today we can find forms that broadly correspond to most of the major evolutionary stages through which life has progressed over the past several billion years. This not only gives us amazing insight into the historical process by which we became what we are today but allows us to interpret the fossils we have in great detail.

Still, none of this means that building up the Tree of Life is easy or that there are no areas in which major questions still persist. Indeed, at the bottom of the tree there is a big question mark. It took a long time for scientists to settle on the basic tripartite division of life-forms into Bacteria, Archaea, and Eukarya that we use today, and even now it's still not clear exactly how these three major groupings are related to one another. Intuitively, you might well think that among them the bacteria and the archaeons would be the most closely related by common ancestry since, as unicellular creatures lacking nuclei, all members of both groups look really similar to the microscopically aided eye. But what if this is a purely primitive resemblance and we should put more weight on the fact that bacteria and eukaryotes share a similar cell membrane structure? Or on the fact that archaeons and eukaryotes have stretches of DNA in their genes that bacteria don't have? This is a real puzzle, and it is largely because we lack what is known as an "outgroup."

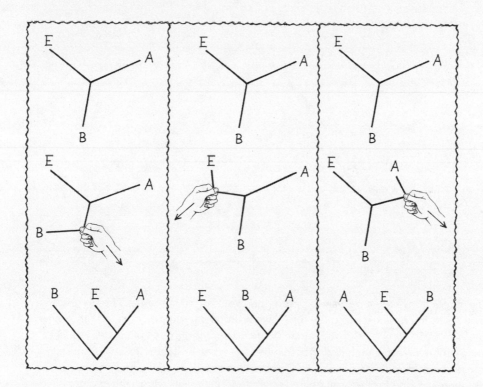

Changing trees by changing the root. The hands are pulling at the branch that will be closest to the root. With a little imagination, the reader should be able to see the topologies of the trees change as the root is pulled.

As we've seen, the principal way in which we can tell which characters we are looking at are primitive, which are derived (and thus informative about ancestry), and which are convergent is through their distribution among the organisms of interest. The best way of demonstrating that a feature is primitive is to show it also occurs *outside* the group you are immediately examining. Unique features, in contrast, are likely to be derived. Outgroups, then, allow us to polarize, or "root," the relationships of these

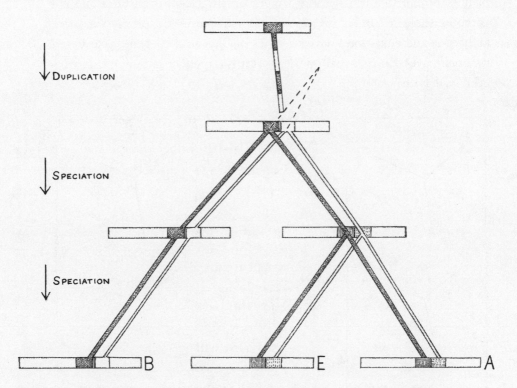

Gene duplications are a pervasive part of the evolution of genomes. The diagram shows a single chromosome at the top. This chromosome gets duplicated (shown as a white box and a gray box) in the common ancestor of the three species A, B, and E. The common ancestor then diverges into two lineages, and then the non-B lineage diverges into two lineages (A and E). The result is two gene phylogenies, one for the dark gene and one for the light gene. The dotted line shows where the two gene phylogenies connect via a common ancestor. In such cases, scientists need to be aware of the existence of multiple gene families. For instance, a study using the dark gene from B and E and the light gene from A would be misleading.

major groups. But unless we find an outgroup on Mars or one lurking in some unexplored spot on our planet, when rooting the entire Tree of Life we are limited to using other, less satisfying means. One possibility is to use DNA sequences of genes that have been duplicated in the genome. When genes are duplicated, there are parallel histories for each of the resulting genes; but since the two kinds of duplicated genes are related to each other, we can put them both in the same tree and observe where the root is in each. When this approach is used, Archaea are shown to be more closely related to us Eukarya than they are to Bacteria. This strange result means that the lack of a nuclear membrane that defines the group Prokaryota, which has been used to embrace both Archaea and Bacteria, is not good derived character, and perhaps we should stop using the term "Prokaryotes" to describe a "real" group of organisms.

The History of a Historical Science

One of the really interesting pursuits in biology is to analyze the distribution of characteristics among the organisms in which we are interested, with the aim of reconstructing what the ancestral forms were like at all of the various branching points in the tree. That way, we can compare known fossils to the reconstructed ancestral forms to see how they might fit into the scheme of relationships. But we are still laboring under the disadvantage that every new feature is only derived at one level in the tree. A derived character state in an ancestor will be primitive in every one of its descendants, no matter how much diversification subsequently occurs. This usually won't leave the poor systematist with too many derived features to consider; and in the case we've just been discussing, since each ancestor must in principle be primitive with respect to all of its descendants and there is no outgroup to appeal to, there may well be no way—even in principle—to reconstruct the exact nature of the ancestral form that gave birth to all life on earth. Such are the realities of science: we have to be prepared to recognize when there is no more we can say, and simply accept it.

Fortunately, there is still a lot that can be confidently said about evolution—and there's a still lot to argue about, too, as there has been from the beginning. When Darwin published *On the Origin of Species* in 1859 there was an immediate outcry as its implications for human origins (which Darwin had specifically omitted) sank into a public mind steeped in the Christian creation myth. But subsequently the notion of humanity's integrated place within the rest of nature came to be accepted with remarkable rapidity. And it remains unchallenged within science, even as the details have become increasingly vigorously debated. Still, in the longer term Darwin's ideas encountered greater opposition not from the public at large but from within science itself, as challenges were voiced not to the basic proposition of descent with modification but to the mechanism of change that he had proposed.

To win acceptance for the notion that life had changed over time, Darwin had to produce an explanation of how such change might have occurred. His clever solution was the theory of "natural selection." This was grounded in the observation that all species are physically variable. Each individual is subtly different from all other members of its species in characteristics that are inherited from its parents. What's more, in every generation more offspring are produced by those parents than can ever survive to become adult and to reproduce themselves. This undisputable fact inevitably leads to a constant process of triage among individuals of each generation, and Darwin proposed that those individuals most successful in the reproductive stakes are those best "adapted" to prevailing circumstances.

Because better-adapted individuals leave more offspring compared to the less well adapted ones who are winnowed out of the reproductive pool, this competition for life and reproduction produces a generation-by-generation alteration in the aspect of each population, as poor adaptations disappear to be replaced by more advantageous innovations. Change thus occurs slowly, via the gradual accumulation of tiny differences over the generations. These eventually sum out to major alterations in evolving lineages. The story is a wonderful one, and clearly there is some truth to it: certainly enough to have made one of Darwin's colleagues berate himself, on reading the *Origin*, with the exclamation, "How very stupid not to have thought of that!" What's more, in the mid-nineteenth century the excruciatingly slow process involved was becoming entirely feasible in light of the hugely long history that geologists of the period were busily attributing to our planet. By the time Darwin was ready to present his ideas to the public, few scientists any longer believed in the six-thousand-year timetable since the Creation that had been calculated from the long succession of "begats" recounted in the book of Genesis.

Nonetheless, with the establishment of the science of genetics at around the beginning of the twentieth century, alternative mechanisms for evolutionary change began to be proposed. Early geneticists noted, for example, that among organisms major innovations were sometimes observed to appear suddenly, de novo, rather than as a result of the slow accretion of tiny changes, and this suggested to some of these researchers that new species might result not from gradual change but from more or less instantaneous genetic events. Debate raged for decades, until, as the second quarter of the twentieth century dawned, mathematical modeling and the application of some powerful intellect led to a convergence that became known as the Evolutionary Synthesis.

In a nutshell, the Synthesis taught that virtually all evolutionary phenomena, even radical modifications, could be traced back to the slow accumulation of small genetic changes over vast periods of time. Even the splitting of lineages was a gradual process, as populations of the same species isolated by geographical barriers gradually diverged down their own separate evolutionary channels, much as a river might divide on encountering a particularly hard outcropping of rock in the middle of its course.

This is a really attractive idea to our reductionist human minds, with all the elegance of simplicity. But after a period of several decades during which it went largely unquestioned, the Synthesis began to run up against the inherent untidiness of nature. The first major attack on the underpinnings of the Synthesis came in the early 1970s, perhaps not surprisingly from paleontologists. Students of the fossil record had been essentially sidelined by the Synthesis; for although the geneticists held the key to evolutionary mechanisms and the systematists were essential for sorting out the great patterns that had emerged from the evolutionary process, paleontologists had been relegated to a fairly humble clerical role in documenting the details of the history of life. However—as Darwin himself had been all too painfully aware—the fossil record in fact reveals vast numbers of discontinuities in that history, and paleontology had in fact been brought in line with the Synthesis by what amounted to a trick: the claim that all these discontinuities were no more than artifacts of the incompleteness of the record. The dissident paleontologists instead made the counterclaim that the persistent gaps in the ever-expanding fossil record might in fact be informative after all. They pointed out that species typically appeared rather abruptly in the fossil record and then lingered around for sometimes extended periods of time (occasionally as much as tens of millions of years) before being replaced abruptly by successor species.

This opened the way for a critical reexamination of the evolutionary process, and it led to some revealing conclusions. Among these, perhaps the most important was that evolutionary change may very often be due to chance events rather than to the genetic fine-tuning involved in the long-term modification of lineages by natural selection.

Keeping the Doctor Away

The evolutionary biologists Stephen Jay Gould and Richard Lewontin pointed out in the 1970s that their contemporaries' perception of evolution at the time was skewed toward looking for and (unfortunately) always finding adaptation. Adaptation is the product of natural selection, which was Darwin's lynchpin for explaining the natural world. And since natural selection was accepted as the great explanatory factor in biology, it was comforting to think that nearly all of what scientists saw in the natural world was the product of that important process. After all, when you see something strange, like a giraffe's neck or a panda's thumb or a horse's toes, it helps to believe that you know the reason they exist. Still, this tendency was unfortunate because it reflected a nonscientific approach to the study of evolution: a narrative way of explaining the natural world that Lewontin and Gould named "just-so" storytelling, in honor of the great storyteller Rudyard Kipling. They pointed out that this line of attack left little room for testing hypotheses and, worse, no room for their rejection. And, as pointed out by the philosopher Karl Popper, modern science rests on the "hypothetico-deductive" process whereby hypotheses are proposed and false ones subsequently rejected.

What is more, as Gould and Lewontin noted, the narrative approach was problematic not only because it was nonscientific but also because it had become paradigmatic. And this, they felt, was harmful both to the expansion of evolutionary theory and to improving explanations of the natural world. They used the Dr. Pangloss character from Molière's play *Candide* to exemplify the problems they saw with just-so storytelling. Pangloss thought there was a reason for absolutely everything, from the Lisbon earthquake to eyeglasses and pants: we have legs so we wear pants, and we have noses so we wear eyeglasses. Having labeled adaptationist narratives the "Panglossian paradigm," Gould and Lewontin suggested that there are four reasons why real-life organisms do not attain perfection via natural selection.

First, novel ways of coping with the environment do not appear de novo as a result of the evolutionary process, like new designs for machines or buildings. The imaginations of inventors and architects are limited not by existing structures but by purpose, and hence a solution to an architectural problem or technical problem can come from anywhere, and in any shape or design. It is true that a lot of technical advances are built on basic advances in science and technology, and that most functional architectural design is based on a general theme (protection from the elements), but the leaps that human invention can make toward technical and architectural perfection far outdo what evolution can do for organisms. Human invention and technology can "step outside the box" and attack a problem from scratch, without worrying about what was there before, but evolution is limited by the hand it is dealt: by the body plans, physiologies, and chemistries of existing organisms. By the nature of the processes that make it possible, evolution cannot stray far from existing organizations and chemistries. These limitations importantly include the way that information is passed on from generation to generation through genes and genomes.

Second, evolution does not always result in adaptation anyway. Some of evolution is just pure good (or bad) luck. It is interesting to note that, at about the same time Gould and Lewontin wrote their famous paper, the importance of "neutral" or nonselected processes in evolution was being explained both theoretically and empirically. Lewontin was himself one of the major players in this arena, especially after he and his colleague Jack Hubby had looked at the gene products in large numbers of the lowly fruit fly. Their goal was to characterize the level of genetic variation in nature, and they expected to find a particular amount of it because prevailing versions of natural selection theory suggested that any variants with less "fitness" than others would be removed from populations. But when they actually did the experiment in the 1960s, they discovered a great deal of genetic variation in their fruit flies that made no sense in the context of pervasive natural selection. As a result, along with other scientists, they suggested that there is actually a large amount of chance, or "drift," in the natural world. Natural selection was not the only bully on the block.

Once this was pointed out, many scientists began to realize that chance events accounted for many of the things we see in nature. And if all those anatomies and behaviors and chemistries were there by chance, then perfection could hardly be the goal—or even the outcome—of evolution. This chance-driven way of evolving might seem strange, but it really is an important part of how the process works. And it happens mostly in small populations, because chance events happen more often when samples are small. Natural selection's twin, if you will, is the process called random drift. Drift happens in small populations because of a phenomenon called sampling error. It can be thought of in the following way. Take four hundred coins and flip them. Although it is possible that you could flip four hundred heads in a row, you would be rather foolish to bet on it (the probability that you would get four hundred heads in a row is one-half raised to the four-hundredth power). The reality of four hundred coin flips is that you would see a lot of heads and a lot of tails (on average two hundred of each). But now take four coins. It would be a reasonable bet that you could flip four heads in a row and not see any tails (the probability of getting four heads in row is one-half raised to the fourth power—that is, you would be successful one in sixteen times you tried). This sampling error principle also works for organisms in natural populations. And it works no matter what kind of organism we are talking about. What drift means with respect to our brains, and the adaptation of our brains, is that much of what we observe is there simply by chance. What we see is the result of random

FLIPPING 100 PENNIES
57 HEADS   43 TAILS

FLIPPING 4 PENNIES
ALL TAILS

If one hundred fair coins are flipped, the outcome will most likely be close to fifty heads and fifty tails, as in the top of the illustration. The probability of flipping all heads or all tails with this many coins is 1 in 1,125,900,000,000,000. If one flips a smaller number of coins (four coins), the probability of getting all tails or all heads is quite high (1 every 16 times you flip them).

events, built mainly on others, with some natural selection tossed in. Not a very efficient way to produce a brain, or any product for that matter. But the critical thing is that it works.

A third reason natural selection does not result in perfect organisms is that, whereas some structures appear to be pretty good if not perfect responses to environmental challenges, they may not come without cost. Possessing them might be penalized by adverse effects in other traits. Take, for example, the strange case of a small Hawaiian fly called *Drosophila heteroneura*. This fly's specific name implies that something is going on with its head; indeed, the male flies in this species have heads shaped like the ends of a hammer, in which their eyes are on the end of thick stalks. Why? Well, it turns out that male flies with really broad heads are more successful at reproducing. Again, why? Two reasons. The first is that female flies simply seem to prefer males with broader and broader heads. In other words, the females find males with large hammer heads "sexier" than the others. The second reason is that males with bigger hammer heads can fight better over females, giving the male a double sexual whammy. But nothing is that simple. The male flies with the broadest heads are likely at a disadvantage because their vision could be impaired by the increasing separation of their eyes. Such an anatomical arrangement could create a visual system with large "blind spots" that make the broader-headed males less likely to see predators coming and therefore less likely to escape predation. If you are eaten, it doesn't matter how sexy you are.

The final reason natural selection doesn't make perfect organisms is that, as we noted, natural selection works on existing variation. The genomes of organisms cannot "conjure up" new variants that might be good at responding to environmental challenges. However well some organisms and viruses may do at adapting to challenges, this adaptation is not the result of consciously creating new variation and using it. For example, although HIV has one of the most rapid rates of response to antiviral challenge in any virus, this is not because the virus itself can sense the kind of changes it needs to make in its genome. Rather, it mutates very rapidly because it is based on RNA, which is notably pliable to mutation as it replicates. Rapid mutation randomly generates genetic variability, which is then acted upon by natural selection. Many bacteria also adapt very rapidly and efficiently to environmental challenges like antibiotics. Again, this is not because the bacteria "sense" that they need variation that can help them develop drug resistance but because plasmids that can carry drug resistance happen to be prevalent in the bacterial world.

To summarize, new innovations arise by spontaneous changes in the genetic material that are entirely unrelated to the ecological trials and tribulations a particular species may be undergoing at any point in time. And although there is no doubt that natural selection is constantly at work within populations, it doesn't necessarily push them in any particular adaptive direction. Instead, it may serve at least equally to keep them in a relatively static state of functional equilibrium, by trimming off the wild extremes of variation. What's more, natural selection can't propel innovations into existence, no matter how advantageous they might be in theory. Genetic innovations arise at random; and where they are not actively disadvantageous, they may survive simply because they don't get in the way or because, occasionally, they may be linked genetically to something else that does provide a benefit. Indeed, new traits often hang around in populations for long periods before, once in a while, they find themselves recruited in some new role—as in the case of feathers, which bird ancestors possessed for millions of years before they ever used them to fly.

Progress in molecular biology has revealed that minor random genetic changes may indeed have major consequences for the physical appearances of their possessors. And it is this phenomenon above all that may well account for many of the morphological discontinuities among fossils that we usually recognize as "gaps" in the fossil record. Finally, it is fair to note that the observable action of natural selection is inherently limited to characteristics that have major survival or reproductive implications. Many stretches of DNA perform multiple functions, and large numbers of what we perceive as independent physical features are in fact closely linked genetically. So, although there is inevitably a strong temptation for us to think of natural selection in terms of physical characteristics whose histories we can trace in the fossil record, selection can in fact only turn thumbs up or down on the survival or reproductive success of entire individuals.

In most cases it will be how you perform as an integrated whole, rather than some key trait you might have, that will determine your success.

On a larger scale, there is no doubt that the greater historical patterns we see in the fossil record cannot be simply the result of within-population changes. Those changes, random or otherwise, certainly provide the raw materials on which evolutionary histories are based, but the grand sweep of the evolutionary drama is provided by the differential survival of species as wholes, by whether or not they give rise to descendant species and by whether those species are themselves successful. Success or failure in the ecological arena will often depend on factors that are entirely external to the actors concerned. Whether you will or will not be able to cope with habitat change depends not only on you yourself but on the amount and the nature of that change, as well as on who is out there predating on you or competing with you for resources. Factors like this are entirely out of your control and have no regard for however magnificently you may have been adapted to earlier circumstances. If an asteroid impact occurs in your vicinity, you will simply be out of luck; if fortune smiles, you may need only to be just good enough to muddle through.

Evolutionary Zigzag

As we've seen, evolution cannot invariably, if ever, be the process of optimization that is implied by the gradual-change model in which species become ever more perfectly adapted to their environments. For one thing, environments tend to change abruptly, on short time scales to which natural selection cannot possibly respond. And if your environment does change suddenly, you are much more likely to move somewhere else more congenial or, if this is impossible, to become extinct than you are to stand still and change on the spot. Indeed, the less tightly you are adapted to an environment, the more flexible you are, the less likely it is that the environment will sock you hard when it changes. There is a price to be paid for optimization, as we see in the fact that extinction rates are much greater among highly specialized species than they are among generalists.

As the product of a long evolutionary process with many zigs and zags, involving ancestors who were living in circumstances that were hugely different from our own, the human brain is an affront to the principles that any self-respecting engineer would follow. And it is in this that its secret lies. For to a large degree, it is the very fact that its history has not optimized it for anything that accounts for the human brain's being the hugely creative and simultaneously both logical and irrational organ that it is.

We are all aware intellectually that human behaviors are sometimes bizarre; the discipline of "evolutionary psychology" has in fact prospered by producing a string of attractive logical explanations for our behavioral failings. The basic idea, born of the simplistic notion of evolution as a process of fine-tuning, is that our brains are somehow out of kilter with the modern world because they were specifically formed by natural selection to cope with an "environment of evolutionary adaptedness" that no longer exists. To evolutionary psychologists, a disconnect between the world we live in and the world we are adapted to occupy accounts for the weird ways in which we sometimes behave. This is a great story, and one with a strong innate appeal to members of what is indeed a storytelling species. But alas—or, rather, very happily—it is just plain wrong. The evolutionary psychologists have certainly played a valuable role in documenting recurring patterns among our behaviors—patterns that most certainly reflect the common structure of the neural equipment residing in the heads of all cognitively normal human beings. But to blame our foibles on now-inappropriately fine-tuned brains is to miss the point entirely. Our brains and the occasionally bizarre behaviors they produce are the product of a lengthy and supremely untidy history: a history that has accidentally resulted in a splendidly eccentric and creative product. Fortunately, the multifarious phases of that saga are reflected in a wide variety of living denizens of the natural world which neuroscientists and others have spent more than a century studying. The rest of this book is an account of the fascinating things they and their paleontological colleagues have learned.

2 THE NITTY-GRITTY OF THE NERVOUS SYSTEM

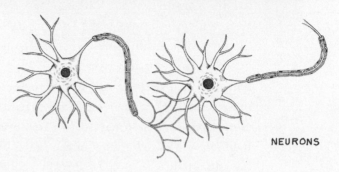

NEURONS

In *The Astonishing Hypothesis*, Sir Francis Crick suggested: "You, your joys and your sorrows, your memories and your ambitions, your sense of personal identity and free will, are in fact no more than the behaviour of a vast assembly of nerve cells and their associated molecules." Actually, one might reasonably find this hypothesis not so astonishing, given that our brains and nervous systems are unquestionably made up of cells, and atoms and ions are indeed the workhorses of the brain. Yet at the same time it certainly seems extraordinary that a mass of electrochemical signals inside the amazing organ residing within our heads can actually add up to our consciousness of ourselves. Surely there is a lot more to us as thinking, feeling individuals than this reductionist account would imply. Isn't this view somehow obscurely demeaning? These two perspectives have coexisted uneasily ever since humans began to wonder about the relationship between body and self—the dichotomy between brains and minds. And to understand this dichotomy better, we need to look closely at Crick's "astonishing" hypothesis and at how our brains actually function. Perhaps the best place to start is with how information is transmitted throughout our nervous systems, a task that requires us to discuss the ins and outs of the universe itself.

It All Started with a Bang

And how! The Big Bang, which occurred at the origin of our universe 13.7 billion years ago, has been the subject of much research and speculation. What we do know is that it happened very fast and that it resulted in the drastic rearrangement of ancient particles into the atoms and elements we see today. Fast was really fast: Big Bang theorists suggest that some of the most basic events resulting in the atoms and nuclei we see today happened in time-space between the 10^{-46} second and the 10^{-11} second after the initial expansion. During this incredibly short time span it was unimaginably hot, so the elementary particles that scientists spend so much effort searching for today could have existed quite easily. But during this fragment of time the mass of the universe was also expanding, allowing the temperature to cool. What really happened at this busy moment is still subject to a lot of debate, but events clarify at about 10^{-11} second, when the energy of particles decreased and the speed of their movement slowed. A short time later (at 10^{-6} second) protons and neutrons, the building blocks of atomic nuclei, were formed. At about 1 second, electrons and positrons formed, though conditions were still incredibly hot. Some estimates place the temperature even a few minutes after the Big Bang at a billion degrees Kelvin (10^9 degrees Celsius). The particles that dominated the universe at this instant were photons: it was still far too hot for any of the protons, neutrons, and electrons to interact with one another. Between one minute and 379,000 years after the Big Bang, the universe continued to expand and to cool. Simple hydrogen nuclei formed, each consisting of a proton and a neutron. The energy from photon radiation was eventually overcome by gravity, and the other forces that dominate the universe today—the weak and strong nuclear forces and electromagnetic force—became predominant. Because these forces that mold atoms came to dominate and the density of the universe was lowered significantly by expansion, electrons could now combine with the hydrogen nuclei to produce hydrogen atoms. Perhaps other combinations of particles and reactive conditions existed, but the stability of the hydrogen atom won the day. At this point, the universe consisted mostly of hydrogen atoms; and in a sense, our nervous system was born at this early point in time.

To oversimplify grossly we can say that, once the hydrogen atom was possible, the universe had begun to develop not only in the context of stellar formation and stellar evolution but also in the context of chemical evolution. Here we want to focus on the runaway train of chemical evolution, because it is this which resulted in other atoms, molecules, and cells. Chemical evolution also resulted in the physical structure of atoms, allowing for electrical charges to be carried by altered atoms called ions. Without ions the electrochemistry that drives our nervous system could not have evolved, and this

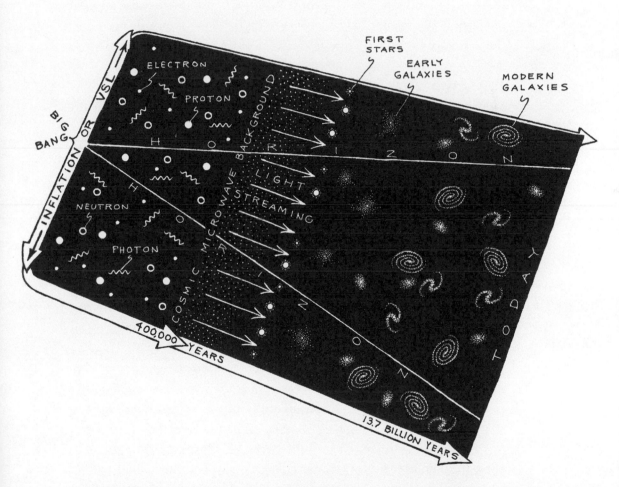

The Big Bang. Most of the interesting stuff for the nervous system happened at the left side of the diagram up to about four hundred thousand years after the Big Bang.

is why we directly owe our existence to the Big Bang and to the chemical evolution that resulted in the following half-million years. As Carl Sagan once said, we are all "star stuff." Our colleague Neil deGrasse Tyson put it another way: "With chemical elements forged over fourteen billion years in the fires of high-mass stars that exploded into space, and with these elements enriching subsequent generations of stars with carbon, oxygen, nitrogen, and other basic ingredients of life itself, we are not just figuratively but literally made of stardust." Moreover, not only are we made of stardust, but this stardust allows life on this planet to move, to emote, and—for us humans—to think and to have consciousness. Both our existence and our realization of it result from the way the original stardust has changed over the past fourteen billion years.

Ions, the Workhorses of Nervous Systems

Chemical evolution is the product of reactions that occurred about half a million years after the Big Bang. These reactions involved the hydrogen atoms formed as a result of the expanding and cooling universe, and they are the outcome of the structure of those atoms. There are many descriptions of the makeup of atoms, but one in particular sticks in our minds. It comes from a popular television show called *WKRP in Cincinnati* that aired back in the 1970s. It involved an African American disc jockey character called Venus Flytrap, a proud and extremely intelligent but misplaced soul at the radio station. In one episode, Venus confronted a gang member to persuade the kid to go back to school. The deal was that, if Venus could teach the gang member the basics of the atom in two minutes, the kid would go to school. Here is a transcript:

> There are three gangs on the street, right? And this (Venus points to a blank wall), this right here is the territory. Now here is the neighborhood (draws a big circle on the wall). Got that? And right in the middle of this neighborhood is a gang called the New Boys (draws a small capital N in the center of the circle [see illustration]). Out here on the outside of the neighborhood, on the edge of the neighborhood, is another gang. You know these are real negative dudes, really nega-

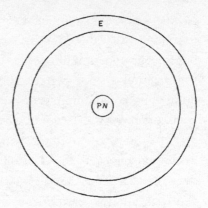

Venus Flytrap's atom on the wall, showing the location of protons (P), neutrons (N), and electrons (E). According to Venus, the nucleus resides in the middle where the PN is, and the electrons (E) orbit around the nucleus.

tive. Right? Now they call themselves the Elected Ones (draws a small capital E on the circle). You got that? Really negative. They don't like nothing. Now they're all the time out here circling around the neighborhood, just circling, checking out the New Boys. Now the new boys see this so they make a deal with another gang. A gang of very happy-go-lucky guys. They call themselves the pros. The Pros (draws a P next to the N in the center). Now the pros are very positive cats. They've got all the good-looking women, right? Now, the Pros and Elected Ones, they hate each other. So much so that they keep the same number of members. Just in case. You dig? You see right here? (points to the P and N). The Pros and the New Boys, they call their hangout the nucleus (draws a small circle around the P and N). Yeah. Now that's a real tough word. It's Latin, and it means center. All right. Give you another word. It's tron. It means dude. All these gangs like that name so well that they all decide to use it. The Pros right here in the middle start calling themselves the protons. And the New Boys, well they start calling themselves the neutrons. And out here on the edge, the Elected Ones, they start calling themselves the electrons. And all this here is the neighborhood (points to the empty area between the center and outer circles). This is block after block of nothing. You understand block after block of nothing don't you?

Needless to say, the kid gets the point. What he will learn when he returns to school is that this atomic structure is actually much more complicated; but for our understanding of how the structure of atoms allows for chemical reactions, it is a perfect description. All stable atoms have, as Venus points out, equal numbers of protons and electrons. The number of electrons and protons determines what element an atom represents. Remember that the universe started out with just a bunch of hydrogen—one proton in the nucleus, and one electron whizzing around the nucleus. How did the atom that Venus describes above form, the one with two electrons, two neutrons, and two protons? This is the helium atom, and its formation billions of years ago is exemplary of how atoms form. Helium atoms started to develop in large quantities when stars like our sun were formed. Because the temperature of such stars is really hot (15,000,000 degrees Kelvin), hydrogen atoms are torn apart to produce nuclei with a proton in them, plus a dismembered electron. So if the helium atom's nucleus has two protons, how did it get them? Sounds simple: just fuse two dismembered hydrogen nuclei, and you'll get a helium nucleus. Well, partly, because we still need to explain where Venus's New Boys, the neutrons, came from. When two protons fuse, they don't simply produce a nucleus with two protons. Instead they create what is called a deuteron, plus a couple of smaller particles. The fusion reaction takes a lot of energy, easily supplied by the incredibly hot temperature of the sun, where this is all happening. But the deuteron produced by the fusion is made of a single proton and a single neutron, so we aren't there yet because we need two of each to make up the helium nucleus. What happens next, then, is that a deuteron fuses with a dismembered hydrogen nucleus (a proton) to form yet another intermediate called a Helium-3 nucleus. This nucleus has two protons and one neutron—still not enough to make Venus's atom with two New Boys, two Pros, and two Elected Ones. But as a result of the incredibly high temperatures where all of this occurs, two Helium-3 nuclei fuse to form a genuine helium nucleus, plus two solitary protons. If this genuine helium nucleus attracts two electrons, it then becomes a helium atom. If not, these nuclei continue to collide and fuse with each other to form larger and larger nuclei that can also attract electrons to form bigger atoms or elements.

There are four basic elements that, because of the way the universe evolved, were involved in making up life on Planet Earth. These are oxygen, carbon, hydrogen, and nitrogen, arranged in the order of their mass in most living things. There are also several other elements that are found in amounts between 0.01 percent and 0.1 percent of total body mass. Some of these, like sulfur and phosphorus, have evolved to be part of organic molecules like proteins (sulfur) and deoxyribonucleic acid or DNA (phosphorus). Others, like magnesium, chlorine, sodium, and potassium, have found a role in the nervous system.

Sometimes the electrons of atoms can be transferred from one atom to an adjacent atom and still remain stable. Such transfer decreases the number of electrons in the donor atom and increases the number of electrons in the acceptor atom. When this happens, the electrical charges of the two atoms change. The donor becomes positive (the happy-go-lucky state that Venus described above) and the acceptor becomes negative (that nasty state). When these transfers happen, the donor and acceptor become what are called "ions." Ions can be pretty stable; when they are, they carry what is called a charge.

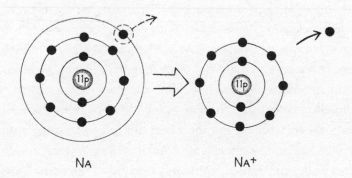

The structure of atoms and ions is dependent on the electrons in the various orbital levels of atoms. A sodium atom is shown on the left with eleven protons (11p) in the nucleus and eleven electrons (black dots) shown orbiting around the nucleus in three different orbits (shown by thin lines). The loss (arrow) of an electron (circled dot in the diagram on the left) results in a positive sodium ion. Gain of an electron (not shown) for some atoms would result in a negative ion of that atom.

One of the most important kinds of reaction that evolved with respect to the brain was the swapping of electrons among elements, such as magnesium, potassium, sodium, and calcium, that make up much less than 1 percent of the atoms in our cells. These ions ultimately became some of the most important elements in our nervous systems. This is because they can easily lose and gain electrons and are in great enough abundance that they can make a difference to how our cells function. In their ionic states these elements can dissolve freely in water, which is what most of our bodies and cells are made of, and they have become the basic currency of our nervous systems. In their ionic states they can move electrons around, and these electrons make up electrical currents, the same kinds of currents that course through the electrical wiring of your house or emanate from the battery of your cell phone. The only differences are in scale and in the particular medium through which the current is moving. For a current to make a big difference, there needs to be an "inside" and an "outside" to the medium through which the ions move. In living organisms, "inside" and "outside" are established by membranes, and this brings us to a big, big question—the origin of cellular life on our planet.

The evolution of atoms and their unique properties is also important for understanding other aspects of the cell. Depending on their unique structures, different elements can combine with others. Such bonding happens in many ways, but often through what is called an ionic bond, which occurs where the two atoms that are bonding actually share an electron. Other kinds of bonds require the input of another element such as hydrogen. The bottom line here is that the structure of each kind of atom encourages particular interactions with the outside world, notably with other atoms. As a result, atoms combine into molecules. Small molecules then combine into longer and bigger molecules, ultimately leading to the complexity of the large biomolecules we observe in cells today. And now that we have completely oversimplified physics and chemistry, we can move on to biology.

On the Origin of Life

Right now there is no firm answer to the Big Question: how life originated on earth. In fact, there probably never will be. Dennis Bray, a biochemist and cell biologist at Cambridge University, put it this way: "Our description of the origins of life is the product of imagination. It has to be—we cannot go back in time to witness the crucial molecular events in the path of evolution. All we can do is to construct plausible scenarios for what might have happened based on the processes that we know do happen now." However bleak that might sound scientifically, it does point to the major goal of science—explanation. We don't have to be entirely right all the time about a natural phenomenon, as long as we can find the best explanation possible. As we've said, "truth" is not a goal in science. In fact, many philosophers feel that the truth can never be attained because we simply wouldn't flat-out know the truth even if we were sitting right next to it. By testing scenarios and ideas about the natural world, however, we can get a better idea of how to explain it. The case of the origin of life is a perfect example of this scientific process. For although we have no real way of knowing if any of the ideas we offer up as explanations is really true, we can eliminate many of them by deducing that they are false. What we end up with is a group of ideas that have survived the testing process. And the scenarios for the origin of life on earth that have survived the testing process are, after all is said and done, pretty good explanations.

For life to have evolved on this planet, several things had to occur to give rise to such complex biological phenomena as cell membranes, the nucleic acids ribonucleic acid (RNA) and deoxyribonucleic acid (DNA), amino acid chains, replication apparatus, nuclear membranes, introns, the Krebs cycle, mitochondria, chloroplasts, and many more: the list could go on and on. This is a lot of things to have happened, though few scientists doubt that the basic structures and processes in the cell are indeed evolvable. Most of the discussion has been about timing: but what is astonishing is how the parts assembled into the whole and the sequence in which this assembly came about.

Specifying this has been so daunting that some highly respected scientists have preferred to push the problem "off planet." The concept of "panspermia" suggests that life on earth did not start from scratch here but was delivered from elsewhere in the universe billions of years ago, on a comet or a meteor. In this scenario, it is most commonly thought that an interstellar or interplanetary microbe entered the earth's atmosphere and seeded all subsequent life on the planet.

Other scientists prefer to start from scratch, proposing a world where the basic building blocks of life exist as a result of simple chemical reactions in a "primordial soup." Molecules then organized themselves from atoms, and a lot of evolutionary "experimentation" resulted in molecules like amino acids and nucleic acids. The continuing formation of larger molecules led to the linking together of nucleic acids and amino acids, and eventually to an RNA world that existed before cellular life. Why RNA and not some other molecule? Well, RNA is a "jack of all trades" molecule. It is made of four basic building blocks arranged in a linear fashion, like beads on a string. One really important aspect of RNA is that these building blocks (called guanine, adenine, cytosine, and uracyl) pair in specific ways. An adenine will always pair with a uracil, and a cytosine with a guanine. These pairing rules mean that if you have a string of nucleic acids, you can easily replicate it. Because RNA can be copied, it can transmit information in a hereditary fashion. RNA can also be used as a scaffold for large, complex structures in the cell, such as the ribosomes. And if that weren't enough, RNA can catalyze chemical reactions, because it can fold into various shapes as a result of its chemical composition.

The RNA world eventually evolved into a DNA-based world, where DNA became the principal purveyor of information. DNA is made up of the same kinds of building blocks arranged in the same linear fashion as RNA (though it uses an amino acid called thymine instead of uracil), so a new language for transmitting information didn't need to evolve. DNA is also a bit more stable than RNA, and because it is really stable in its double-stranded form, it proved to be an incredibly efficient way to copy and transmit information. Next in the sequence of events, proteins became the workhorses of cells, because of their greater functional efficiency and plasticity. Proteins are made of amino acids, again arranged like beads on a string; and because twenty different amino acids are used in forming proteins, the potential for diversity among proteins is much greater than among nucleic acids. In addition, the long protein chains (like nucleic acids) can fold to produce myriad three-dimensional forms.

How cells go from the four-letter code of nucleic acids to the twenty-letter code of amino acids is the central theme of the genetic code. RNA is still important and essential in the DNA world, with lots of functions in the cell. But in the post-RNA world, a major driving force is natural selection. The self-organization and cooperation of the tiny molecular machines is a highly advantageous trait for both replication and survival. But in order for the chemical constituents of cells to self-organize and cooperate, they need to be in close contact with one another, as well as prevented from cooperating with other machines. This is where membranes come in. Whether membranes came before or after simple molecular machines like RNA is not that important to our understanding of the nervous system, but the simple fact that membranes formed in the first place is critical to the story.

The formation of membranes seems to have been a fairly simple process, once lipids had formed. Lipids are long chains of atoms made up mostly of carbon and hydrogen, and they compose about 40 percent of all the organic matter in our bodies (that is, ignoring the water and salts). Lipid chains can be very long, and they hate water. This makes them insoluble in water and perfect for forming barriers to it. Membranes contain a particular class of lipids that differ from other lipids such as saturated and unsaturated fatty acids by the addition of a "phosphate group" that yields a "phospholipid" chain.

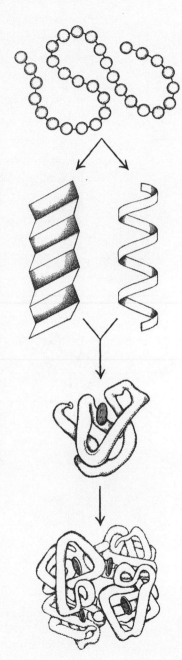

The four levels of complexity of a protein. The Primary level (top) is simply the linear sequence of the building blocks of proteins called amino acids. The Secondary level (second from top) is the transformation of the linear array of amino acids into a nonlinear structure, usually a helix or a sheet. The Tertiary (third from top) level is the folding of the helices and sheets into a three-dimensional structure. The Quaternary level (bottom) is the complexing of subunits into more elaborate multi-subunit entities.

In contrast to the surrounding lipids, the phosphate group loves water. In its absence the lipid molecules will just float around, bumping into one another and going about their business of being carbon and hydrogen atoms linked together. But if you put a lot of phospholipids in water they will organize themselves according to their likes and dislikes. Since the lipid ends of the phospholipids hate water so much, they will arrange themselves away from water. But the only way they can protect themselves from it is to try to sit next to another phospholipid chain, "kissing" the other chain where the lipid parts are and shoving the phosphate part out into the water. Because the phosphate end of the phospholipid likes water, it doesn't mind this. And the series of interactions produces what is called a lipid bilayer, with two phospholipid chains zipped together at their lipid ends and their phosphate ends pointing out into the water so that the lipid

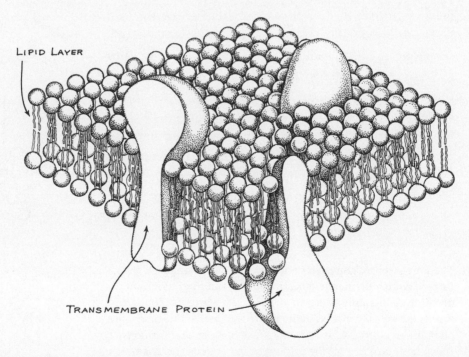

The lipid bilayer, showing transmembrane proteins. The balls in the lipid layer are the hydrophilic heads of the phospholipids with the hairpin tails hanging off. The tails of the phospholipids are hydrophobic.

ends can hide from it. From this propensity it is easy to see how, with enough phospholipid around, you could readily produce a balloon with a lipid bilayer skin that more or less isolates the inside from the outside. And this is, of course, just what is needed to contain the ions and molecular machines important for life within cells and to hurry along the evolutionary process for cellular life and nervous systems. Also, once this is accomplished, the arrangement of lipids is important for implementing ionic or chemical gradients between the outside and the inside.

There is one final aspect of membranes that we need to discuss before we move on to things electrical. Membranes are fairly fluid structures. They are like loosely woven fabric, into which it is easy to stick objects. In this case the objects are "membrane proteins." These proteins readily fit within the matrix of the membrane and perform many tasks important to the maintenance of ionic and chemical gradients inside and outside the cell. They also implement many other cellular functions, including immune tasks, cell-to-cell communication, and transport of nutrients across the cell membrane. And they are an important part of the story of how the nervous system works.

Electric Boogie

One way to think about electrical currents in your body is to think about how electricity works in your house. The wires themselves are made of metal, which has large numbers of free electrons sitting around; and the electricity coursing through those wires is nothing more than a stream of free electrons. But of course the electron streams don't appear out of nowhere. If nothing is applied to the metal wire, the electrons will just wander aimlessly. To get the electrons moving in a specific direction, you need to apply a negative charge to one end of the wire and a positive charge to the other. When this is done, the electrons sitting around in the metal of the wire start to move, specifically toward the end of the wire where the positive charge is applied. The scatter of the electrons moving through the wire is broad, and that is why most of the wires in your house have plastic around them. Plastic doesn't conduct electricity, so when a wire is covered with plastic and electrical forces are applied to its ends, the electrons will move more smoothly and universally toward the positive end of the wire. The plastic thus serves as an insulator; this concept will become important a little later, when we talk about the insulation of the neural projections we call axons by a substance called myelin.

Some metals conduct electricity better than others, and the current can also be held back by the phenomenon known as resistance. Basically, the more free electrons there are present, the lower will be the resistance. Electrical current is measured by estimating the absolute electrical force in volts and dividing by the resistance. Simple but elegant, this calculation gives you the basic quantity of electrons that will course through a wire of given resistance after the application of a given electrical force. The current is measured in units called amperes.

Apparently, the minimum electrical current that the human body can detect is about 0.001 ampere (A). What this means is that any current greater than this will be detected by our brains as odd, and sometimes as painful. A current of 0.1 A can be lethal to humans if the current passes through a sensitive part of the body like the heart or the brain. Currents as low as 0.01 A have been known to cause fibrillation (severe alteration of the heartbeat). This happens because when the electrical current starts coursing through the body the stream of electrons confuses and disguises the normal electron movements among our cells, causing our muscles to do things in exaggerated ways. If the outside current is moving through heart muscle, then bang, there goes your heart. How much voltage does it take to shock us? Again, this depends on the resistance, and when we are exposed to electrical forces in the guise of volts, our skin acts as resistance. We get badly shocked when the voltage is above 200 V, the point where the skin no longer can impede (resist) the electrical current.

As it happens, our neural cells communicate with electricity just like the wires in the walls of your house. So why aren't we buzzing all the time? The answer is actually that we *are* buzzing all of the time, just not at a high enough electrical current to register a signal of pain or discomfort in our brain. Instead, the currents that course through our cells deliver important information to our brains without causing us to register their passage. But when a large enough current is applied to our bodies, it causes the normal flow of electrons in our cells to fritz out.

A normal electrical impulse in our nervous system is tiny compared even to the small force coursing through your cell phone. Scientists who study the action of electrical forces in the brain call such impulses "potentials." Instead of measuring potentials in amperes, they are presented in volts. A cell acts a lot like a battery, with positive and negative poles that correspond to the inside and outside of the cell, depending on what state it is in. When a cell is just sitting there with no current, it is said to be resting. But even though the cell is resting, there is still a difference in the concentration of the electrons on one side of the cell wall versus the other. This difference is caused by the disparity in concentration of the various ions that exist inside and outside of the cell. For instance, the extracellular concentration of sodium (which is a positive ion and therefore lacks electrons) is about ten times greater than the intracellular concentration of sodium. This means that, with respect to sodium, the extracellular area of a neuron is positive, while the intracellular region is negative in the resting state. On the other hand, potassium, which is also a positive ion, is found in higher concentrations on the inside of the cell than on the outside. And chlorine, a negative ion, has the same pattern as sodium, higher outside than inside. The bottom line here is that a resting cell has a voltage of −70 millivolts (mV). This is minuscule considering that our wall plugs have 120,000 mV flowing out, and even the typical AA battery can move out 1,500 mV when connected up.

The electrical signals that move up and down our nerve cells occur when the resting potential is changed by altering the concentration of electrons on the inside and outside of the neural cell. The change in charge that takes place when this occurs is from −70 mV to +30 mV, and it's accomplished by moving charged ions back and forth across the cell membrane. Neurobiologists call this change an "action potential." Because the cell wants desperately to be in its grounded resting state, it will return to −70 mV. This movement of ions across the cell membrane is achieved using proteins called "channels" that have properties like swinging gates. We will discuss later how proteins can behave like machines; for now, consider a gated channel as a kind of swinging gate. Because sodium and potassium ions are involved, and sodium likes to go into the inner side of the membrane whereas potassium likes to run out to the outer side of the membrane, these channels are directional. A sodium channel can only allow sodium through the gate to the inner side of the membrane, and a potassium channel can only let potassium through the gate to the outer side of the membrane.

Here is how an action potential occurs across an ideal membrane (voltages for different kinds of membranes vary). In the resting phase both kinds of gated channels (sodium- and potassium-gated channels) in the

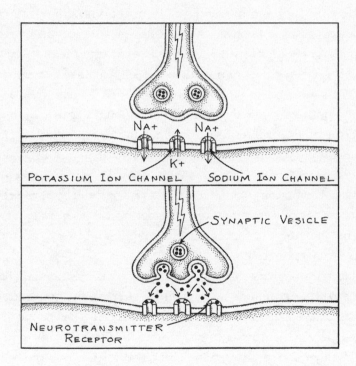

A synapse. This diagram shows how an action potential is converted into a molecular signal at the synapse (top panel) by release of neurotransmitters from the synaptic vesicles (bottom panel). The action potential causes the fusion of the vesicle with the presynaptic membrane, and release of the neurotransmitter occurs. The neurotransmitters bind to the ion channel receptors (open circles in the bottom figure), and this results in the opening and closing of the ion channel.

membrane are closed. As the result of a charge in the membrane caused, say, by communication with another cell, the sodium-gated channel will open when the voltage on the inside of the membrane is −50 mV. When the sodium-gated channel opens, it lets sodium in until the voltage of the membrane rises to +30 mV, at which point the membrane has an action potential. Remember that sodium is a positive ion, so the voltage will go up from the resting −70 mV. When the potential near the localized region of the membrane where all of this is happening reaches +30 mV, the sodium channel closes and the potassium-gated channel opens. When the potassium gate is open, potassium—also a positive ion—rushes through to the outer side of the cell membrane. This lowers the voltage across the membrane until it reaches −70 mV, at which point the cell has returned to its ground state, or resting potential. This all happens in about 1.5 milliseconds, or 0.0015 of a second.

Because the cell membrane is lined with these ion-gated channels, a chain reaction can be initiated and maintained along a cell membrane. If the cell is round and blobby, the action potential will spread like a wave

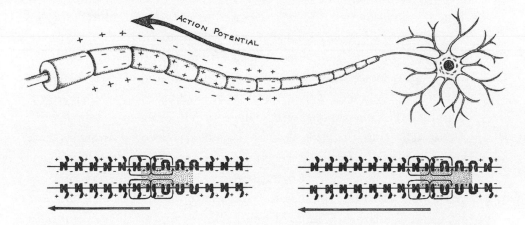

A neuron, showing the movement of an action potential across an axon. Below the neuron, diagrams show the ligand gated channels opening and closing, causing the action potential to move along the axon. The diagram on the left occurs before the one on the right. The symbol + indicates a positive charge, and − indicates a negative charge. The boxes around the channels along the bottom are where the gated ion channels are changing to produce the action potential.

over the membrane of the blob. But if the cell has a linear or wirelike shape, the chain reaction will cause the action potential to travel in a line. It turns out that neural cell structure may be both blobby (the soma or cell body) and wirelike (the axon). But it will not be entirely wirelike. Just as the electrical wires in our walls of our houses are insulated with plastic or rubber, our nervous system evolved specialized cells that act like insulators. These cells are called glial cells, and they form an interesting alternating pattern along the axon that increases efficiency in the movement of action potentials along the arms of the neural cell. And at this point, before we discuss the basic structure of nerve cells in more detail, we need to describe how a single genome can give rise to the large number of different kinds of cells that exist among organisms on this planet.

The Evolution of Cell Types

The diverse functionality of animal bodies is mostly due to the hundreds of different cell types that have evolved among them. As we will see, some animals like sponges have very few cell types and hence a pretty limited functionality compared to primates like us that have hundreds of cell types. What makes one cell a different type from another is simply the proteins it contains. It is these that make different structures in the cells, or that implement different functions in them; and they are coded for by genes in the genomes of the cells. If we are talking about a single organism, then the same genome will be in every cell, producing the many different cell types that exist in that organism. Stem-cell scientists have recently begun to understand the genetic mechanisms of cell-type determination, but how cell-type diversity evolved remains a mystery. As European Molecular Biology Laboratory researcher Detlev Arendt put it: "Cell types are fundamental units of multicellular life, but their evolution is obscure." Nonetheless, Arendt has expended considerable effort in trying to understand the evolution of cell types to define some "first principles of cell type diversification."

How and what genes are regulated in a cell are part of what is called a "molecular fingerprint." These molecular fingerprints are the principal tools employed to understand the origin of different animal cell types. In most cases, it is possible to determine what gene or genes are responsible for major kinds of differentiation of cells; these genes become very important for establishing a molecular fingerprint. What's more, really small RNA molecules have been implicated in the control of gene expression in cells. These micro-RNAs, or miRNAs, can be followed in cells, and the kinds and quantities of them are indicative of cell type. Once a molecular fingerprint database has been set up, discovering how cell types are related to one another is simply a detective story, much like CSI.

Central to understanding the brain is the question, Where did nerve cells come from? Arendt suggests that they arose from ancient multifunctional cell types. When we examine very simple animals (chapter 3) with limited numbers of cell types, we can observe that cells which get specialized in more complex animals (with more cell types) serve multiple functions. For instance, in the cells of some simple animals, structures for epithelial muscles, myofibers, cilia, secretory vesicles, and glandular functions are composed of a single cell type. The diversification and establishment of multiple cell types in more complex species would necessarily have involved the formation of sister cells, each with a separate function. Sister cells can diverge in several ways. One is by the splitting of function, with one cell specializing in one part of the original function and the sister cell in another. Another is by separating into sister cells with similar but slightly different functions. A third is by one sister cell acquiring a completely new function. Such segregation of function appears to be the major

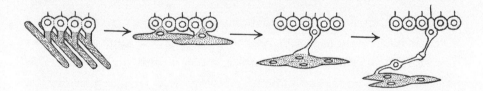

Two cell segregation events lead to the evolution of a neural cell. The cells on the far left are myoepithelial cells. These segregate into the two intermediate cell types (two figures in the middle) and eventually lead to the neural cell arrangement on the far right. After Arendt 2008.

route by which ancestral cell types with multiple functions evolved into neural cells. Events like these go a long way to explain how our nervous system became so diverse and how simpler nervous systems can give rise to more and more complex ones.

Opening Pandora's Box

In the blockbuster movie *Avatar*, writer and director James Cameron created Pandora, a world in which communication among and between species is a key to the survival of life. Cameron's premise is interesting and well researched; apparently he consulted University of California at Riverside botanist Judie Holt about the kinds of plants that might exist on such a world and how these plants might "behave." Cameron went so far as to consult on the latinized names that certain plants on Pandora might have. But perhaps the most interesting aspect of life on Pandora concerns how the plants and other organisms communicate with one another. Played by Sigourney Weaver, the scientist Grace Augustine claims that the plants are communicating by some form of "signal transduction." Not bad: as we will see in chapter 3, although plants don't necessarily communicate cell-to-cell using the same electrochemical system as animals, they do produce electrical currents. And they do indeed use signal transduction.

If James Cameron had made *Avatar* in the 1970s, he most likely would not have used signal transduction as a flashy term to describe the communication between organisms on Pandora, because in 1975 only two or three scientific papers had ever used the phrase in conjunction with cells. Indeed, this is a term most often associated with physics and electronics. But contrast the 1975 situation with the more than ten thousand scientific papers that had made reference to signal transduction in a cellular context by 2005. This increase coincided with the revolution in genetics, cell biology, and molecular biology that occurred in the last two decades of the twentieth century. And an even sharper spike in the number of papers on cellular signal transduction coincided with the development of genomics technology at the turn of the century. The more genomic and cell biology information we obtain, the more we learn about how cells communicate with one another. It is not that the process of signal transduction is new; scientists knew that it existed. But, as with *Avatar*, it needed a catchy name to be accepted as a fundamental cellular process.

There are several components to signal transduction. It starts with the reaction of a receptor molecule, usually embedded in the membrane of the cell, to a molecule or stimulus from outside the cell. This then *transduces* a response by the cell. Such responses can be anywhere from slight to extreme—and they are sometimes spectacular. The receptor molecules or proteins are usually much more complex molecules than the ones stimulating them from the outside. In fact, most molecules that initiate signal transduction are really small. These tiny molecules are also called ligands, and they basically float around between the cells as a result of the everyday cellular lifestyle. But when they come into contact with a cell membrane, they bump into receptors that recognize them. Ligands actually fit physically into molecular pockets on the receptors. This hand-in-glove capacity is possible because the receptor proteins have a three-dimensional structure. They can make long strands or big, globule-like structures and everything in between.

Molecules in the cell can also act like machines, and intricate ones at that: current models of how proteins work reveal that there are different makes and models of molecular machines in our cells. For instance, a molecule called "ATP synthase" acts like a turbine to change the conformation of other proteins along its "shaft" that, in turn, implement the conversion of a molecule called ADP to ATP. The conversion of ADP to ATP is a fundamental reaction in the cell and is basic to many cell functions. A molecule called kinesin, which transports large molecules about the cell, literally walks along other molecules called microtubules to carry its cargo to different and specific parts of the cell. Yet another molecular machine is the ribosome, a collection of many protein and nucleic acid components. The ribosome takes a nucleic acid, manipulates it, reads it, and makes proteins using the nucleic acid as a script. Molecules like these have many of the properties of our man-made machines; more important, they mostly have properties of machines we have never heard of—nor would ever think we might need. Machinelike molecules can bend and break structures; they can move about; and they can move other molecular components around. They can connect to other molecules, and when they do so, they change shape. This is very important, since the shape of a molecule in many cases determines its function. So if a small molecule collides with a big one and changes shape as a result, its function might change. Molecules can also interlock; in doing so, two molecules can essentially become one and attain a new function that wouldn't exist if they were not connected. Molecules can also be affected by outside forces like electromagnetic waves or sound waves, a property that brings us to such senses as seeing and hearing, which we will discuss in chapter 4.

Some of the more commonly recognized signal transduction components are hormones. Everyone has heard of these molecules: they are those crazy messengers that float around our bodies and for which we blame such things as sexual tension and teen misbehavior. Hormones are small molecules that are secreted by different clumps of cells, organs, or glands in our bodies and travel through the bloodstream. One of the more important in terms of sheer survival is the hormone called adrenaline (epinephrine), which is secreted into the bloodstream as a result of physical injury, exercise, cold, and emotional shock. Just about anything that unexpectedly upsets us—or any vertebrate, for that matter—will cause a surge of adrenaline into the bloodstream, to warn the rest of the body that action is needed. Some hormones, like estrogen and testosterone, belong to a class of molecules called steroids, and the molecules with which they interact are called steroid receptors. Steroids will course through the bloodstream, bumping into cells until they get transported inside one, where they will encounter and bind to steroid receptors. The part of the steroid receptor molecule that recognizes the steroid is known as the ligand binding domain. Another part of the protein, called the DNA binding domain, does the signaling to the cell after the steroid has bound to the receptor protein. Binding to the steroid activates the transport of the receptor to the nucleus, where it interacts with the genome of the cell.

At this point, the DNA binding domain of the activated steroid receptor is just itching to do what its name implies—bind to DNA in the nucleus.

And when it does so, the nuclear steroid receptor affects how the regions of DNA that it binds to are expressed into other proteins. In essence, the nuclear steroid receptor is signaled by the hormone to regulate the expression of other genes that then respond to the initial presence of the steroid hormone.

Another category of receptor proteins important in signal transduction, and even more important to our understanding of the brain, is the neurotransmitter receptors. Neurotransmitters, like hormones, are small molecules. In fact, many hormones are also considered to be neurotransmitters. One of the smaller and more important neurotransmitters is an amino acid (yes, the same kind of small molecule that makes up proteins) called glutamate. This neurotransmitter is extremely important to the nervous system, because an excess of it will trigger certain neurological responses. Unlike nuclear steroid receptors that interact with steroids after they enter the cell, most neurotransmitter receptors are embedded in the cell membrane and wait for a neurotransmitter like glutamate to bump into them. This all happens in the "synapse," where two neurons "air kiss." When neurotransmitter and receptor collide, the neurotransmitter binds to what is called an active site on the receptor, and the conformation of the receptor changes. In most cases this causes a gate to be opened to the inside of the cell. There are well over a hundred different neurotransmitters with which our nerve cells deal, and there are even more neurotransmitter receptors that implement the binding of these intracellular molecules to the cells of the synapse.

With all of these ligands, hormones, neurotransmitters, receptors, and other molecules floating around, how do our brains and the rest of our bodies know what to do? It turns out that everything is not as chaotic as it might seem. Only certain glands, or patches of cells, will produce a specific hormone or transmitter. At the other end, many areas of the brain specialize in only one or a few kinds of receptors. And some neurotransmitters will interact with more than one receptor, thus expanding the kinds of signals than can be transduced by neurotransmission of a single kind of molecule. By enhanced specialization of tissues and parts of the brain, some order is imparted to this rather complex system. For instance, the neurotransmitter known as gamma aminobutyric acid, or GABA, interacts with two major kinds of receptors. The first kind is called a ligand gated ion channel, involving the kind of protein we discussed previously that implements the movement of ions across the membranes of nerve cells. These are fast-acting receptors that, upon the binding of GABA, implement the opening of a gated channel. Such receptors are also known as ionotropic receptors. The fast-acting GABA receptors inhibit the firing of an action potential, so GABA acts like a sort of fast-acting tranquilizer.

The second kind of GABA receptor is called a metabotropic receptor. It acts quite differently, by using what neurobiologists call a second messenger protein. These receptors implement a slow response and employ a category of intracellular proteins called G-coupled proteins. The GABA binds to the GABA metabotropic receptor that is embedded in the cell membrane. This interaction changes the shape of the GABA receptor, which then travels to the inside of the cell. Once there, it interacts with a G-coupled protein that implements a slow-acting response in the nerve cell.

Signal transduction can also induce complex interactions in the cell that result in very specific and intricate control over cell function. One category of neurotransmitter receptor proteins that shuttles information to the cell from the outside is known as G-protein-coupled receptors, or GPCR. The "G-protein" in the name refers to intracellular proteins that act like molecular switches. These are made up of distinct subunits, each of which has a function when on its own, plus a separate and equally important function when it is complexed together in the G-protein.

The GPCR has a unique structure. Remember that proteins are linear affairs, with a starting end and a butt end, and that they can fold into interesting shapes and forms. The GPCR protein folds upon itself such that there are seven rather rigid rods of amino acids called helices, with six short loops connecting them. The helices are arranged parallel to each other in the membrane, but the connector loops are so small that the helices have to be arranged head to tail, where the head is the part of the helix closer to the starting part of the protein and the tail is closer to the butt end of the protein. If we look at the protein from left to right as it is embedded in the membrane, we also see that the start of the protein hangs off the first helix, while off the seventh and final helix is a little tail of amino acids. This makes a total of eight little elements of GPCRs that need to be tucked away somewhere. And as a result, GPCRs are integrated into the cell membrane so that four of these little elements flop out of the cell membrane to end up outside, and four flop toward the inside of the cell. GPCRs are a diverse array of proteins connected by this characteristic seven-helix structure. There are almost a thousand GPCR genes in the human genome, and the function of only about 20 percent of them is understood so far.

These proteins implement the communication of the cell with the outside world in many ways. GPCRs have two major methods of doing this: the "cyclic AMP" and the "phosphatidylinositol signal" pathways. These

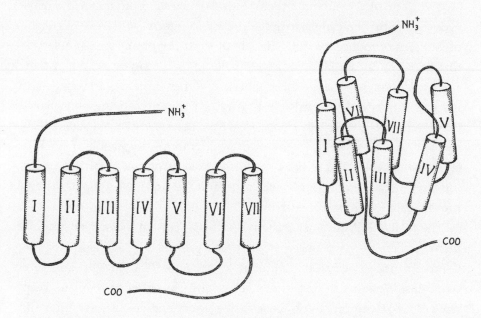

A typical G-protein-coupled receptor. The seven cylinders represent the seven domains in the protein that reside in the cell membrane. The lines represent those parts of the protein that reside on the outside of the cell (top) and the inside of the cell (bottom). The amino terminus of the protein is labeled NH3+, and the carboxy end of the protein is labeled COO. The diagram on the left shows a GPCR spread out, while the right-side diagram shows how the intramembrane domains are arranged in the membrane.

work in a stunningly mechanical fashion. To make a long story short, an external stimulus such as a hormone or a neurotransmitter binds to the GPCR embedded in the membrane. This binding causes a change in conformation of the GPCR, and the positions of those four loop elements that are sticking out into the inside of the cell change subtly. But even subtle changes like this are very attractive to other proteins in the cell; indeed, a G-protein will interact with the subtly altered GPCR, and the G-protein will then bind to a molecule called guanosine triphosphate (GTP). The binding of GTP to a G-protein causes it to dissociate into its different subunits. So far, so good. But these subunits alone will have little impact on the cell itself, even though the reason why GPCR binds to the neurotransmitter in the first place is to implement a change in the state of the cell. But even though just causing the G-protein to dissociate doesn't do much on its own, one of the subunits then binds to another protein called adenyl cyclase, and this binding causes the cyclase to interact with ATP (a cousin of GTP) and produce cyclic AMP, which is an important small molecule that then regulates a kind of protein called protein kinase A. When they are active, protein kinases "phosphorylate" other proteins, a process that inactivates or severely retards the activity of the protein being phosphorylated. In this way, the chain reaction we have described can regulate the activity of various proteins in the cell. What is more, the whole process is entirely reversible. This is because all the proteins involved will snap back into their initial conformations when the stimulus stops, whereupon the entire "machine" will simply reset itself.

The cyclic AMP, or cAMP, system is called a "second messenger system" because there are two major steps in the process: the activation of the G-protein and the activation of protein kinase A, or PKA. But why are there two steps? Wouldn't a "first messenger system" be easier to regulate, keep track of, and fix if something went wrong? Every engineer knows, after all, that the more moving parts in a device, the more likely it is that something will break. Well, there are two answers to this question. First, as with anything in nature, the GPCR pathway wasn't designed. Instead it evolved, and thus was not optimized. Second, though, maybe it isn't that inefficient after all. Specifically, using the second messenger cAMP has two huge advantages—

amplification and speed. Cyclic AMP is very easily produced by the process we've described. Little energy is expended in producing it, which makes it a cheap and efficient way to amplify the number of second messengers that then implement the activation of PKA, responsible for cell regulation. And it's fast because cAMP is a small molecule that can diffuse throughout the cell easily and can be manufactured very rapidly.

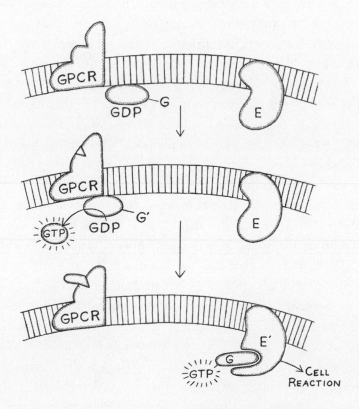

Second messenger system. A G-protein-coupled receptor (GPCR) embedded in the membrane is coupled to a G-protein (labeled by G), with a guanosine diphosphate (GDP) molecule bound to it. The GPCR interacts with a ligand (small triangle, in this case a hormone), and this leads to several reactions in the cell. First, GDP is converted to guanosine triphosphate (GTP), which then causes the release of one of the subunits (alpha) of the G-protein. The resultant altered G protein can then bind to adenylate cyclase (E), resulting in the triggering of further reactions in the cell.

Mind the Gap! The Connectome Is About to Arrive

So far so good, but how does communication between cells work? How does the signal of the action potential "jump" from one cell to another? To understand this important form of cell-to-cell communication, we need to delve a bit deeper into the microstructure of neural cells and how they "air kiss" each other.

The road to understanding nerve cells and their structure was actually a pretty rocky one. It involved two histologists who scrutinized each other's work to the point of persecution. Ironically, although they disagreed terribly with each other, the two scientists in the feud, Camillo Golgi and Santiago Ramón y Cajal, eventually shared the 1906 Nobel Prize for their work on the nervous system. The prize-winning research was made possible by Golgi's invention of a stain (named after him) that allowed researchers to view only neural cells. Although scientists before them had understood that the brain and nervous system were made up of cells, Golgi's stain allowed them for the first time to visualize them.

The disagreement arose after this point. Golgi thought he had proven the reticular theory, an old idea that claimed the entire nervous system was made up of a single solid network of fibers caused by the fusion of neural cells. Cajal improved Golgi's staining technique and came to a completely different conclusion. By examining the histological preparations from his lab, Cajal concluded that the nervous systems of organisms were made up of large numbers of individual cells, and not of a single network of fibers. This became known as the "neuron doctrine," and Cajal proposed that the nerve cell tips (which Golgi had ignored) were somehow in contact with each other but that each cell retained its autonomy. When they went to Stockholm to receive their shared Nobel Prize, Golgi completely ignored Cajal's work in his acceptance speech. Very awkward! As Cajal said: "What a cruel irony of fate . . . like Siamese twins united by the shoulders, scientific adversaries of such contrasting character!" We now know that Cajal's ideas about the cellular nervous system with the cells communicating by means of some kind of contact is correct and that the reticular hypothesis was wrong. But it took Cajal a long time to convince the scientific community of the veracity of his interpretation.

What Cajal was actually seeing at the cell tips were the "synapses," the key to how nerve cells communicate. A basic nerve cell has four components. The major part of the nerve cell, where its nucleus and genome reside, is called the cell body, or "soma." "Dendrites" are the bushy or treelike projections that come out of the cell body, and they are where most of the signals from other cells are received. The third nerve cell component is the "axon," a long projection that emerges from the cell body and reaches toward other nerve cells. It is coated with "glial" cells to insulate it. The last part of the nerve cell is the "axonal terminal," which is where the cell forms synapses with other cells. Nerve cells come in a wide variety of sizes and shapes, depending on their location and function in the nervous system. And they are also classified according to their function and direction of communication. "Afferent" or "sensory" neurons are those that transmit information from the outer world to the brain. We will have plenty to say about the senses in chapter 4. There are also the "efferent" or "motor" neurons that transmit information away from the brain to parts of our bodies, and induce organs or muscles to perform work. And then there is everything else, grouped as "interneurons." These communicate with other neural cells within the nervous system.

The longest nerve cell documented is found in the aptly named colossal squid. Nerve cells in these animals can be up to forty feet long. It's possible that the blue whale, the largest living vertebrate, has even longer nerve cells, but this isn't known for sure. While we're quoting amazing figures, we should note that the brain of a human being has around 100 billion cells at birth. Other organisms like dogs (1 billion) have fewer, because their brains are smaller. Mice have only 75 million. Some of the organisms that we will shortly discuss in connection with the evolution of senses have far fewer neurons yet. For instance, the fruit fly *Drosophila melanogaster* has 100,000. *Aplysia californicus*, the California sea slug that has been used in neurobiology studies of memory, has 20,000. And finally, the nematode *Caenorhabditis elegans*, which is made up of a little over 1,000 cells total, has 302 neural cells.

The number of connections for a typical nerve cell is 1,000. This means that on average our brains have $1,000 \times 100,000,000,000$ or 100,000,000,000,000 (100 trillion) connections. And in fact, because all that is needed to make a neural connection is a little bit of dendritic real estate, the number of places synapses can form is theoretically much greater even than this 100 trillion number. Using the area of axons as a guide, the number of potential connections in a human brain has been estimated to be more than 10 to the 76th power. That is a 1 with 76 zeros after it, a number in the same range as the number of particles in the known universe. These connections among nerve cells are what the neurologist Charles Scott Sherrington named synapses, and it is their particular biology that makes our nervous system so flexible. As we've outlined, much of the activity of the nervous system is electrical, pulsing on through the activity of the action potential. But there is also a lot of chemical communication taking place among proteins and neurotransmitters and receptors. And how this chemical communication works is also essential to our understanding of the nervous system.

How are all of these nerve cells and synapses arranged in the brain? The MIT scientist Sebastian Seung and his colleagues have started a huge project to understand how neural cells are related to one another and where they interconnect. In the spirit of large-scale projects such as the Human Genome Project (all of the genes in the human cell) and the Human Proteome (all of the proteins produced by the human cell), Seung and his colleagues have coined another "ome," this time the connectome, to designate the project to decipher the connections within brains. As a first step in this project, Seung and his colleagues have produced a colorful, fascinating, and informative way of viewing the brain. Their approach is to start with a 1 mm × 1 mm cube of mouse brain (about the size of the head of a pin). They fix the cube, cut from the brain, in chemicals that stop the activities of the cells in the tiny cube but preserve the cells' structures. They then make thousands of tiny slices of the cube and place them in an electron microscope to view the slices. Now, because most neural cells are longer than they are wide, what you see in a typical cube are the axonal projections of the cells that course through the cube. By following the

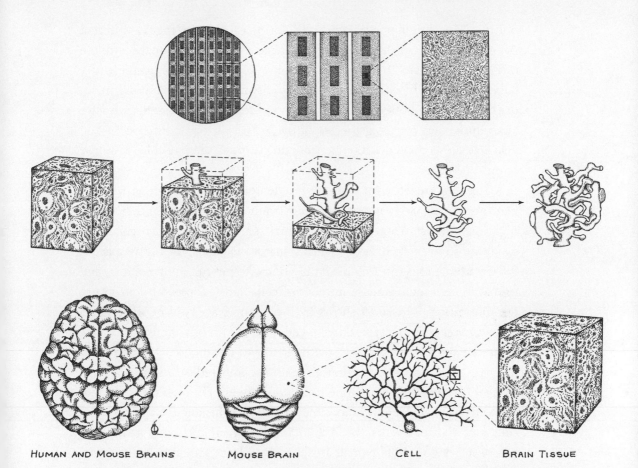

HUMAN AND MOUSE BRAINS   MOUSE BRAIN   CELL   BRAIN TISSUE

Sebastian Seung and colleagues' "connectome." The series of diagrams on the top shows the general process by which the data are obtained. Sections of a block of brain tissue from a mouse are obtained and then observed in the electron microscope. These sections are sequentially arranged so that a computer can recognize the same cell in each sequential micrograph. In the middle series of pictures the process of reconstructing the neural cells is diagrammed. As each section of the sequence of sections is analyzed it is removed, and the resulting nerve cell axon and dendrites are revealed. The bottom series of diagrams shows the scale at which the analysis is accomplished. The block of tissue is shown on the right. The next diagram to the left is of a nerve cell, with its typical treelike appearance. A typical nerve cell in a mouse brain (shown in the third diagram from the right) would take up the space marked on the mouse brain. The leftmost diagram shows a mouse brain drawn next to an adult human brain.

outlines of each cell in each slice and coloring each one of them a different color, Seung and colleagues are able to make a picture of a square with the outlines of each cell in a particular color. When this is done to the entire cube, they are left with hundreds of little squares that correspond in sequence from the bottom of the cube to the top. Using massive computer algorithms, they piece together all of the squares in three dimensions and visualize the axons and cell bodies coursing through the tiny cube of mouse brain.

What they are left with is an unprecedented three-dimensional view of neurons passing one another, winding around one another, and "air kissing" one another. The three-dimensional structures they have generated so far are stunning and show that in some of the cubes, the entire space is filled with neurons running in all directions. To put this all in context, and to drive home the immense number of cells and synapses in your brain, the illustration on page 59 shows by increasing scale how big that 1 mm cube is compared to a typical nerve cell in the brain of a mouse. Next, we show the size of the nerve cell with respect to the mouse's brain, and finally we show how big the mouse's brain is compared to a human brain.

In this chapter we have gone from nothingness, through the Big Bang, to electrons to atoms to ions to molecules to proteins and, finally, to nerve cells and synapses. It has been a 13.5-billion-year journey starting well before the origin of life, which arose on this planet about 3.5 billion years ago. Many of the mechanisms that determine how molecules interact with one another, and with cells, started their 3.5-billion-year history with the origin of tiny organisms that vaguely resembled modern-day bacteria. Subsequently, many of the ways in which our nerve cells work began to be honed with the origin of "eukaryotic" cells possessing membrane-bound nuclei. So next, by hanging our brains on the Tree of Life, we hope to uncover some of the complexities that were involved in the long and tortuous journey to our modern human brains.

3 HANGING OUR BRAINS ON THE TREE OF LIFE

SLIME MOLD LIFE CYCLE

To understand the long and winding route leading up to our modern brains, we need to understand how cells work in other organisms. This is no meager requirement, because the variety of organisms on this planet is daunting. But scientists have made a lot of progress toward classifying animals and, hence, toward making understanding the story possible. To get there, though, we need to examine the Tree of Life, because the general genetic conservatism of organisms means that a range of organisms alive today approximates many of the stages through which a succession of ancestors may have passed on the long road to getting here. So in trying to grasp how our extraordinary human brain evolved, it is important to understand the nuances of how other organisms communicate and store and manipulate information.

From the smallest, seemingly most insignificant bacterium or archaeon to the larger and more complex organisms on earth like us, there is an unbroken chain of ancestry and descent. By knowing the who, what, when, where, and how of these ancestors scattered throughout the Tree of Life, we can trace how our brains and those of other organisms have evolved: the patterns we uncover help unpack the history of brain evolution.

Solving this puzzle is similar to tackling a Rubik's cube. To do it, we have to make a large number of moves. There is only one solution (all sides end up the same color), but there are many, many ways to reach it. There is a best way to solve the problem, namely the quickest way; this would be the perfect solution. But however we solve this puzzle, we can make only one set of moves, though everyone's set might well be different. If we wanted two people to solve it in exactly the same way, the second one would need clues as to what the first one did. And so it is with the Tree of Life, in which reconstructing our ancient extinct ancestors provides the clues that help us figure out how the evolutionary story most likely happened. The solution we uncover more than likely won't have been the perfect one, for it turns out that the routes that evolution takes are usually convoluted.

Do We Have a Quorum Yet?

Life on earth is made up of three major "domains" of living organisms. These are the Bacteria, Archaea, and Eukarya (eukaryotes). Bacteria, at least, should be pretty straightforward. They are single-celled and may make us sick, though most of the time they simply coexist with us. It is, however, surprising and sobering to realize that only seven thousand species of bacteria have actually been named (most of them of medical importance), though perhaps as many as ten million (if not more) bacterial species exist on the planet, in all kinds of environments and ecological niches.

Archaea is an even more elusive group of organisms. As the name implies, they have a mystical ancientness to them. They were named for archaic lifestyles that scientists felt might mimic the lifestyles of the most ancient forms of life on our planet. And they are unique today because of their habits of living in really nasty environments with high salt concentrations, extreme pHs (both highly acidic and very alkaline environments), high pressures, or hot temperatures. For this reason they are called "extremophiles." The third big domain of life is Eukarya. These are a very diverse array of organisms, and are distinguished from all other life on the planet by the presence of a nuclear membrane inside the cell (which will become important when we delve deeper into communication within and among cells). We are eukaryotes, and so are most of the organisms on this planet with which we are familiar. But in terms of species or individuals, there are at least an order of magnitude more Bacteria and Archaea on this planet than Eukarya. During the past decade, thousands of Archaea and Bacteria have had their genomes sequenced, and some amazing discoveries have emerged. Focusing on microbes has revealed that single-celled organisms exist in all three of the major domains of life. In the Bacteria and Archaea, every member is single-celled; among Eukarya, the only members that are not single-celled are the animals and plants.

We can make a good argument that any multiple-celled organism needs to have an intricate system of communication among cells. This communication can be for almost any purpose—for example, telling the next-door cell to stop growing, or to grow a different way, or even to commit suicide. Such communication requires constant transfer of information between cells. How is this achieved? Well, when it comes to communication, cells deal in the currency of molecules, atoms, and particles such as electrons. And all three of these have huge impacts on the neural cells that make up animal nervous systems (chapter 2).

It is perhaps harder to imagine that Bacteria, Archaea, and simple eukaryotes such as amoebas need to communicate with one another. These organisms are usually considered the lone wolves of life. They are single-celled, and they were widely supposed to be solitary and to like it that way. In fact, the study of communication among bacteria was rather rare before the past decade, since the convention was that bacterial cells were pretty much self-contained. How wrong can you be! Recent research has turned up many examples of bacteria communicating with one another, and there are in fact many functional and evolutionary reasons why they should communicate. Individuals not only communicate within species but can also communicate across large evolutionary distances. Obviously, their communication is not through sound or sight, as ours is. Instead, it is through molecules.

One of the better-known ways that bacteria communicate is through a process called "quorum sensing." This is a funny but illuminating choice of terminology. The term implies that bacterial cells sense the presence of the many other cells and then react to this large body of cells with some response. Now, bacteria obviously don't have brains, and it might seem that "sensing" is an odd choice of word for the phenomenon. Or is it? Actually, the mechanisms of sensing in bacteria are all molecular, and sensing in complex eukaryotes may also be accomplished at the molecular level. So, throughout the Tree of Life, molecules turn out to be at the heart of sensing changes.

Quorum sensing among bacteria was first discovered in bioluminescent forms related to the cholera bacterium *Vibrio*. Two species of *Vibrio* that live in the ocean luminesce only if enough of them are present. One of these bacteria—*Vibrio fischeri*—has been studied extensively because of the interesting places where it does its "sensing." It lives in tiny sacs that exist in a variety of host organisms. The relationship is "mutualist," which simply means that both organisms—in this case, the host and the bacterium—gain benefit from the interaction. The little sacs are called "light organs," and they serve as tiny culturing chambers for *V. fischeri*. Now, producing light is an energy-intensive operation, and it does no good to luminesce as a lone bacterium: no other organism will see the minuscule amount of light you produce, so this expenditure of energy would not be productive. On

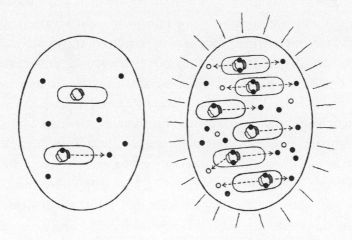

Quorum sensing. The bacteria on the left are producing a signal, but there are too few bacteria for the signal to be useful. The bacteria on the right are producing the same signal, and because they have reached a "quorum," the signal is effective.

the other hand, if there are enough bacteria present in a sufficiently small space, making light through bioluminescence really has a huge visual impact on any organism that might be looking toward the light organ. So *V. fischeri* uses a quorum-sensing mechanism to figure out whether it is inside the light organ or outside. If it is inside, then it will start to make light by luminescing. If it is outside, it will not, because it won't do any good: the bacteria need to be in high concentration in a small area for the light each one makes to add up to a visual impact. This trick is a pretty good one, because the ability to know inside from outside without a brain is actually rather amazing.

How does *V. fischeri* do this? Each bacterium makes a hormone that promotes its own production. If it is in sufficient concentration, the luminescence response occurs, and the *Vibrio* then pump out high volumes of the hormone. If they are in a light organ, the hormone accumulates, and production of it is further amplified. But if they are outside the light organ, the hormone will diffuse, its concentration will be low, and no autoinduction will occur. It turns out that the number of *V. fischeri* needed to make light inside a light organ is about 100 billion cells. The major quorum sensing system at work in *Vibrio* is found in other bacteria, too, and at least four other quorum sensing systems exist independent of the one in *V. fischeri*, involving similar numbers of bacteria. Here we have a system where sensing of a position in space is accomplished by a molecular mechanism, without any brain involved; remarkably enough, we will see that most neural functions do basically the same thing. What's more, the genomes of bacteria also respond vigorously to the potential of communicating with other cells: thus about 10 percent of all of the genes in the common bacterium *Pseudomonas aeruginosa* are under the control of cell-to-cell communication systems like that of *V. fischeri*.

Quorum sensing is also thought to be involved in community interactions of large numbers of bacterial species. As noted, bacteria were long thought of as solitary. They are often observed as free-swimming and unassociated; recently, though, the association of bacteria into biofilms has opened a new way of thinking about bacterial community interactions. What's more, quorum sensing can also act across species boundaries, and there have even been experimental instances of quorum sensing working across kingdoms of organisms. As Melissa Miller and Bonnie Basler of Princeton University put it: "Quorum sensing allows a population of individuals to coordinate global behavior and thus act as a multi-cellular unit." Hence, the evolution of multicellularity in eukaryotes is our next step in understanding how cells communicate.

Does Dicty Deem?

Multicellularity has evolved independently at least twenty times on this planet. Although this number of evolutionary events may not sound like much, it does indicate that, over the long haul, evolving a multicellular lifestyle is relatively easy. Important for beginning our discussion are three kinds of single-celled eukaryotes: ciliates, myxomycetes, and cellular slime molds. These are the lowliest of eukaryotes, but as we will soon find out they have overcome their lowliness with some very elegant ways of doing things.

Ciliates are a category of single-celled eukaryotes that, as their name implies, have cilia: tiny appendages with which they move and feed. Members of some species of ciliates make contact with one another for various reasons, one of which is, not surprisingly, for sex. Actually, their "talking" to each other is strictly molecular. It works a little like our immune system and hence is called "quasi-immune." Our immune systems evolved to determine self from nonself, and the ciliate mating communication systems evolved to allow any two ciliates to determine if mating would be proper. The system works by way of membrane-bound and membrane-associated receptors that bind to proteins secreted from a potential mate. If molecular interaction takes place, then contact and mating can proceed—a simple mating "ritual," with gifts and "yes" meaning yes. Some scientists have even called the mate-seeking molecules secreted by ciliates "pheromones," chemical signals that are an important aspect of mate recognition in many animal species. But although it does seem reasonable to call these molecular interactions in some way pheromonal, we must also keep in mind that ciliates have no nervous system, so their response is very different from what we see in the so-called higher organisms. Still, many of the mechanisms that ciliates use for cell-to-cell recognition have also been found in multi-cellular eukaryotes.

Have you ever been at a diner or restaurant where they have placemats for kids with games on them? Usually a crayon or a pencil is provided, and there are several little games on the mats to keep the children amused. One of the better and more attention-getting games of this kind is the maze game, where the kid has to use a pencil to trace the best way out of a maze. Usually, it's not as easy as it seems. But there are single-celled organisms that are actually pretty good at mazes. These are myxomycetes (also called "true slime molds," a not-so-appealing name), another category of single-celled eukaryotic organisms that communicate with one another across cell membranes. The slime mold has an amoeboid shape that can form tubule-like structures to connect to food sources. And it has been called a "computing machine," because it can assess diffusion patterns among molecules around it and make "decisions" about where to project tubules for food. It can actually quite efficiently solve problems involving mazes. But can we call these organisms "smart?" If we define smart as being able to respond successfully to a challenge that takes intelligence for another organism to deal with, then true slime molds such as *Physarum polycephalum* could actually be called smart. But we have to be careful. Some humans are good at solving mazes. And clearly our smartness at this task emanates from a completely different set of mechanisms than those used by a true slime mold. The slime mold's approach is strictly algorithmic, molecular, chemical, and machinelike. But as we will see, perhaps our ways of doing things are algorithmic, molecular, and chemical, too.

The last single-celled eukaryotes we need to look at here are the "cellular slime molds," which sounds only marginally better than "true" slime molds. These are really interesting and are most notably represented by a fruiting cellular slime mold called *Dictyostelium discoideum*, or Dicty for short. They have a Dr. Jekyll and Mr. Hyde existence, for most of which they live in the soil as single-celled amoeboid-like organisms looking for food and grazing solitarily on soil bacteria. But when Dicty is starving, it congregates into a multicellular mass. This congregation has been likened to a social club, and hence some scientists have suggested that these organisms

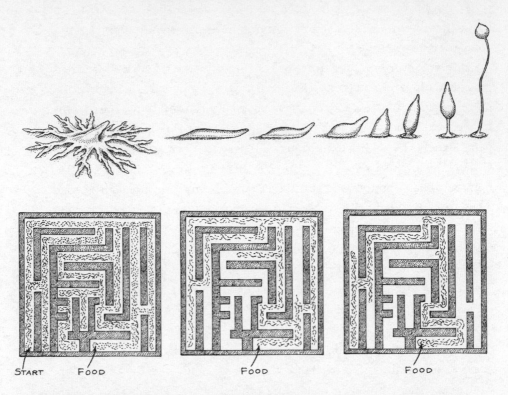

The life cycle of the slime mold *Dictyostelium,* top, and the slime mold *Physarum* "playing" a maze game, bottom. The maze on the left shows the first phase of how *Physarum* "find" the food source. As the "game" goes on, the *Physarum* only grow and extend into the maze parts that are most closely connected to the food source. All routes but the shortest are eliminated by the growing slime mold.

are actually social. Indeed, this lowly slime mold has also been called the "social slime mold" and has been used as a model for studying cooperation and other aspects of sociality. But here is the really interesting phenomenon. Dicty forms a wormlike "creature" made up of thousands of individual cells, each capable of locomotion, and the mass can make "decisions" to go in various directions, much like true slime molds in mazes. In fact, the social slime mold is also really good at solving mazes. And as if that weren't interesting enough, these single-celled organisms have sophisticated mechanisms for cell-to-cell communication. Indeed, they need to communicate in order to take on that wormlike form. What's more, a large number of signaling processes in this lowly single-celled organism actually resemble pathways in our own cells. These signaling pathways are molecular and help the slime mold to communicate both with other cells and with the external environment. And even though this single-celled organism doesn't have a brain or even a nervous system, it has become an important model for understanding neural disorders. This is because Dicty has mitochondria, those powerhouses of the cell. All mitochondria have a small genome that codes for proteins, and when the Dicty mitochondrial DNA is mutated in certain places, the worm stage is prevented from navigating mazes and the poor thing becomes quite confused. What is most amazing of all is that the mitochondrial genes that are disrupted in Dicty to cause this confused state are the very same genes that, when disrupted in humans, cause neural disorders. Does Dicty deem? Does it think?

Once more, we can look at this question from several perspectives. Because the social slime mold doesn't have a brain or a nervous system, Dicty clearly doesn't deem the same way we higher animals do. But this doesn't mean that Dicty isn't doing something with gene products that behave, in their molecular and cellular functions, just like those in higher animals. Thirty percent of the genes in the Dicty genome are also found in the human genome, and some of these are clearly important in nervous system function in higher animals.

Do Plants Have Brains?

Some people think that plants respond to talking and other forms of human attention. And although plants more than likely do not process human language, they are nonetheless very aware of their surroundings and are very capable of communication among their cells. Furthermore, some scientists think the system involved is very close to what we could legitimately call a nervous system. After all, mimosas are famous for retracting rapidly after being disturbed, and Venus flytraps react swiftly to the presence of insects in their capture devices. Charles Darwin made comparable observations and proposed similar ideas about plants. In one of his less well known works, *The Power of Movement in Plants*, he wrote about the radicle, the embryonic root in a plant: "It is hardly an exaggeration to say that the tip of the radicle thus endowed, and having the power of directing the movements of the adjoining parts, acts like the *brain* of one of the lower animals; the *brain* being seated within the anterior end of the body, receiving impressions from the sense-organs, and directing the several movements" (our emphasis). Darwin was saying that not only does the radicle behave like a brain by directing the functions of other cells but it is also positioned in the right place in the anatomy of the plant. Modern botanists have extended this idea. In 2005, the first international plant neurobiology meeting was held in Florence, Italy, and a brand-new journal, *Plant Signalling and Behavior*, was launched in 2006. Just what are the plant neurobiologists proposing?

The idea that plants have nervous systems stems from several sources of information. First, plants have genes that are similar to those that specify animal nervous systems. Such genes include glutamate receptors, neurotransmitter pathway activators such as G-box proteins, and a protein called "14-3-3." All these proteins are found in plants and observed in animals, in which they have been shown to have distinct roles in neural function. Second, although the functions of these proteins more than likely do not have "neural" origins in plants, some plant proteins do behave in a way very similar to neurotransmitters and other neural proteins. Third, some plants do show synapselike regions between cells, across which neurotransmitter molecules facilitate cell-to-cell communication. Included in the requirement for comparison is that the regions should have the same characteristics as animal synapses, such as vesicle formation. Fourth, plants have vascular systems that can act as conduits for the "impulses" that they need to transmit to the rest of the body of the plant. Last, plants have action potentials, as do animal neural cells.

Let's look at these various kinds of information and at what they may imply for the existence of brainlike functions in plants. The first point concerns the presence in plants of genes that are related to genes in animals involved in the nervous system. Actually, this is hardly surprising, and confirmation of this fact was one of the first really interesting results of the various genome projects. The reason why it isn't surprising, of course, is that all life on the planet is united through common ancestry. To find genes in common among broadly divergent organisms is what you'd expect with descent from common ancestors. Thus a typical bacterial genome turns out to have the equivalent of 7 percent or so of its genes in the human genome. For plants the number is about 15 percent, and for organisms like flies and worms the number jumps to 30 percent. For vertebrates, the number is about 85 percent for our most distant relatives such as fish, to 98.7 percent for our closest living relative, the chimpanzee, and 99.7 percent for our close extinct relative, *Homo neanderthalensis*. What was not so expected, though, is the broad distribution of major gene categories that are represented in both plants and animals.

Still, the evolutionary process can facilitate some remarkable processes with genes. A gene that makes a protein involved in a particular process in plants doesn't necessarily have to make a protein that has the same function in an animal or a fungus. For instance, glutamate receptors are involved in the animal neural synapse and interact with the neurotransmitter glutamate. In animals these receptors are found primarily in synaptic regions of nerve cells, and they implement glutamate-mediated postsynaptic excitation of neural cells—which is a fancy way of saying that glutamate is involved in exciting the cell across a synapse. Glutamate is an amino acid, and it needs to be transported across the synapse in order to implement the postsynaptic impulse. It happens that two major kinds of glutamate receptors are recognized on the basis of how they transmit the postsynaptic impulse. The first kind is ionotropic, meaning that glutamate receptors lining the ion channel pore at the synapse activate the pore upon binding of the receptor to glutamate. Metabotropic receptors indirectly activate ion channels through signaling cascades that are usually linked to G-proteins. To allow the whole process to happen, the receptors also have to bind what are called agonists. There are three major kinds of agonists that interact with ionotropic glutamate receptors: AMPA (α-amino-3-hydroxyl-5-methyl-4-isoxazole-propionate), NMDA (N-methyl-D-aspartic acid), and kainate.

There are several versions of the glutamate receptors for both the ionotropic and metabotropic functions, as well as several within those functional categories that are specific for AMPA, NMDA, and kainate. So there are multiple versions of genes for the proteins in animals (this is what is called a gene family). For instance, mice and humans have four ionotropic glutamate receptors that use AMPA as an agonist, seven that use NMDA as the agonist, and five that use kainate as the agonist. Likewise, mice and humans have eight metabotropic glutamate receptors that each use a variety of agonists. Plants have glutamate receptors that are more similar to the ionotropic kind. *Arabidopsis thaliana*, that workhorse of plant genetics and genomics, has twenty members of this gene family. What is curious is that there is a similar number (sixteen) of ionotropic glutamate receptors in animals.

Do the animal ionotropic glutamate receptors have a one-to-one relationship with the plant receptors? In other words, are the four animal glutamate receptors that use AMPA as an agonist more closely related to any four of the plant glutamate receptors than to any other animal or plant receptors? Although it appears that, via duplications in common ancestors, animals have all evolved the same genes in this gene family and hence do have one-to-one relationships, plant glutamate receptors all appear to be evolved from a single common ancestor that existed *before* plants and animals diverged. This means that the very specific glutamate receptors of animals do *not* have a one-to-one relationship with the plant glutamate receptors. The complexity of the animal nervous system use of glutamate receptors would, then, bear little resemblance to the plant use of the receptors.

It is true that plant glutamate receptors can interfere with animal glutamate receptors, suggesting that the plant receptors still have some equivalent function in animal nerve cells. There is, for instance, the strange case of human ingestion of cycads (plants rich in glutamate) causing Alzheimer's-like symptoms. And the expression of plant glutamate receptors is specific to the root, the very location to which most scientists supportive of plant nervous systems point. But although three major categories of plant glutamate receptors have been discovered in plants, echoing the three major categories of ionotropic animal glutamate receptors (AMPA, NMDA, and kainite), the three categories of plant glutamate receptors bear no resemblance at all to these animal categories. Further, although there are three major categories of plant receptors, they do not show distinct organ-specific expression profiles. And while a small subset does appear to be important in early development of the roots of plants, the conclusion is that these proteins bear little functional resemblance to animal glutamate receptors. Still, if glutamate receptors don't serve nervous system functions in plants, why are they there? The most common argument for their retention in plants is that they serve as defense proteins to ward off invading insect species.

Given all this, how do we explain the presence of plant structures that behave like synapses, along with molecules that behave like neurotransmitters active in the "synaptic" region? For this to mean anything, a few characteristics of plants need to be confirmed. Synaptic communication must be shown, implemented through neurotransmitters and through neural transmitter receptors in the same way as in animal neurotransmission—for example, through vesicles near the synapse. So, are there plant molecules that might act like neurotransmitters? One neurotransmitter candidate is auxin (indole-3-acetic acid), a small molecule that some botanists feel is the best argument for a neurological behavior in plants. There are also transporters for auxin that behave a lot like receptors. But when you combine the two, does the auxin system act like a neural reaction? Some scientists would actually argue yes. Botanist Gerd Jürgens at the Max Planck Institute for Developmental Biology, for example, has shown that auxin transport is accomplished through vesicle trafficking that has animal neurotransmitter–like features.

Still, auxin is not found in animals, and it appears to be a plant-specific protein that regulates growth. To some, Jürgens's observations suggest that the vesicle structures might be similar enough to make a good argument, and when the kinds of "synapses" made in plants are examined, two junction types turn out to have membrane-encased protein domains in them. The one we have just been looking at is influenced by light and gravity to control cell-to-cell communication and uses auxin as a transmitter, behaving in the same way as a neurotransmitter. The other synapse behaves like an animal immune-cell synapse does with pathogenic cells. In animals, this system implements the immune response and the destruction of the invading pathogen. In plants, it allows the individual not only to deal with pathogens but also to stabilize interactions with symbionts: an important function. Plants establish useful two-way interactions with a lot of microbes such as bacteria and fungi, and in some cases they accomplish tasks that the plant is unable to do on its own. Some plants cannot process environmental nitrogen, so they form a symbiotic relationship with bacteria from the genus *Rhizobium* to do the trick. In the process, the rhizobia get the benefit of being fed by the plant.

What about electrical impulses in plants? Oddly enough, such electrical impulses were discovered nearly a decade before Luigi Galvani did his ghastly 1780s frog-leg experiments showing electrical impulses in animals. So there is no doubt that electrical signals or action potentials exist in plants. It is also pretty clear that, as Eric Davies at North Carolina State University put it, "the fundamental reason plants have electrical signals is that they permit very rapid and systemic information transmission so that the entire plant is informed almost instantly even though only one region may have been perturbed." Still, the nature of the action potential is quite different in plants and animals. Whereas animals produce the action potential by sodium-potassium exchange, plant potentials are produced with calcium transport that is enhanced by chloride and reduced by potassium.

So what do we conclude? The notion that plants have brains in some sense is both interesting and thought-provoking. So provocative, indeed, that in 2007 investigators from thirty-three institutions published an open letter in a widely read plant sciences journal stating that they "maintain that plant neurobiology does not add to our understanding of plant physiology, plant cell biology or signaling," and imploring the proponents of the initiative to just "cut it out." Overall, the response from the plant neurobiologists on the matter of plant "brains" has been rather conflicted. Anthony Trewavas of the University of Edinburgh suggested that "plant neurobiology is a metaphor"—and nothing more. His focus was on the term itself. And his interest was principally in its importance in driving science to understand the cell biology of plants, as well as on the significance of the interesting and provocative observations made about plant cell-to-cell communication and signaling. But the European biologists František Baluška and Stefano Mancuso strenuously argued for the literal existence of nervous systems in plants, suggesting that "removing the old Aristotelian schism between plants and animals will unify all multicellular organisms under one conceptual umbrella."

Obviously, both perspectives cannot be right. And the argument for metaphor seems to us to be the more reasonable one, given the scientific and evolutionary context of the phenomena used to support the existence of brains, or even just of synapses, in plants. As we noted in chapter 1, the theme of confused sameness ("convergence" or, as Darwin put it, "analogy") is pervasive in the study of the nervous system. Trewavas seems to us to call it like it is: simply a case of discussing similarities. It is the metaphor itself that makes statements about the sameness of plant and animal systems so interesting. But to make it useful, you have to acknowledge that it is metaphor. If we do unify plants and animals under a single "conceptual umbrella" when there really isn't one, then we will have a genuine problem. For one thing, there is good evidence that plants and animals do not share a common ancestor to the exclusion of all other organisms on the planet. Fungi and many single-celled eukaryotes get in the way. And the unifying umbrella would both disguise this and obviate the utility of the metaphor. When a metaphor is no longer recognized as such, fallacy becomes the rule of the day.

Sneezing Sponges, Fact or Fiction?

Speaking of metaphors, University of Alberta invertebrate zoologists Glen Elliott and Sally Leys use an excellent metaphor to suggest that sponges can sneeze. Sponges make their livelihood by filtering water via internal choanoctye chambers. The water is pumped into the chambers by flagellae that beat under the sponges' control. Too much sediment or junk in the water is problematic for sponges, and the evolutionary process has facilitated mechanisms for dealing with the dirt. There are three major kinds of sponges, each with many species. These three major groups are the Desmospongiae, Calcarea, and Hexactinellidae, and the mechanisms for dealing with sediment have evolved differently in each. For the cellular sponges (Desmospongiae and Calcarea), the regulation occurs through contractions and compressions of the chambers. The glass sponges (Hexactinellidae), which are probably unable to contract or compress, get rid of the sediment by stopping their flagella from beating to control water flow, using calcium action potentials.

So two very different ways of controlling an important feeding behavior have evolved in sponges. By putting ink in the general area of water ingestion in a cellular sponge, Elliot and Leys observed a "sneezing" response that caused noxious or harmful materials to jet away from the sponge. Sponge biologists look at this phenomenon as a behavior, asking how the "sneeze" is controlled. If it is implemented in the same way that higher animals mediate their behaviors—that is, through a neural response— then the case might be made that sponges have a nervous system. However, it appears that sneezing in sponges is probably not produced by changing electrical impulses. Instead, the neurophysiologist Robert Meech calls the sponge sneeze "non-neural coordination." In addition, the contractions that cause the sneezing are very deliberately timed and are not instantaneous. As a result, we have to conclude that one of the most conspicuous of sponge behaviors lacks a neural basis. But it does turn out that some sponges do use electrical impulses to modulate rhythmic body movements such as contractions and undulations and, more important, that the sponge genome

A glass sponge contracting and expelling ink in a coordinated fashion. To feed, sponges filter water via internal chambers called choanocytes. The water is pumped into the chambers by flagellae that beat under the sponge's control. Too much sediment or junk in the water is problematic for sponges, and so the evolutionary process has facilitated a mechanism for dealing with the dirt: sponge sneezing. A sponge sneeze takes several minutes to develop and be executed.

contains many genes making the same proteins that are active in the synaptic systems of higher animals.

The only other animal group without nerve cells is one of the most neglected animals of all time. This is a tiny marine animal in a phylum (Placozoa) that is made up entirely of a single species called *Trichoplax adherens*. Actually, work done in 2010 indicates that there are many more than one species of this interesting animal, but because they are so simple, they all look pretty much alike on the surface and are hard to tell apart. Placozoans are transparent little sheets of several thousand cells that actually have a top and a bottom. Their simplicity resides in the fact that among these thousands of cells there are only four cell types, none of them neural. As a standard of comparison, sponges have ten to twenty cell types, depending on the type of sponge. Flies have more than ninety cell types, and humans have over two hundred. Still, Placozoa actually stalk and eat prey, locomoting by cilia or by amoeba-like movement. They can detect whether food is present, and if it is, a distinctive behavior occurs. The tiny sheet of cells that makes up each animal flattens out and scarcely moves. If food is not detected, the sheet moves rapidly and randomly. This is not exactly sneezing, but it is behavior, and indeed quite a conspicuous behavior for such a simple form. Yet as in the sponges, no electrical impulse is involved.

How the Placozoa are related to other animals is controversial, and we will address this question after we talk about some other animals at the end of this chapter, because knowing this is pivotal to understanding both what a nervous system is and what a synapse is. But meanwhile, why is this understanding so important? And if getting an accurate Tree of Life is really needed, what can we use as an outgroup for animals? After all, we made a

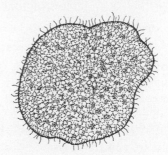

The lowly placozoan *Trichoplax adhaerens* is a transparent little sheet of several thousand cells. In this organism, there are only four cell types, none of which is neural.

big fuss about outgroups in chapter 1, and clearly what is required is an outgroup that is close to animals but not actually an animal. We could use plants, but frankly they are too distantly related to make a good outgroup in this instance. We could use a slime mold (deeming Dicty might be used), but Dicty is actually farther from animals than fungi.

After a bit of hunting around, it turns out that the best outgroup for animals would actually be two groups—choanoflagellates and fungi. Believe it or not, fungi are more closely related to animals than they are to plants, and so they are well positioned to serve in this role. Choanoflagellates are single-celled organisms that form filter-feeding multicellular colonies. And they are commonly recognized as the closest organism we know to animals without actually being one. Using both sets of organisms as outgroups helps ensure that we can accurately interpret the information we have about animal synapses and nervous systems. In other words, using fungi and choanoflagellates as benchmarks should clarify whether synapse and nervous system similarities are derived (inherited from a common ancestor), primitive, or convergent. Interpreting the nervous systems of other organisms in this way tells us a lot about our own.

Is a Neural Net a Nervous System?

Sponges and placozoans are at the base of the animal part of the Tree of Life. The next players in this story are bizarre animals by any criterion we might wish to choose. They are all marine creatures, grouped under the name Cnidaria. In fact, Cnidaria form their own phylum, just like the single-species phylum Placozoa and the other animal phylum we've discussed so far, the sponges (Porifera). Cnidaria have about a dozen cell types, among which is an interesting one named cnidocytes. This odd name comes from the Greek word *cnida*, which means nettle, reflecting the fact that cnidocytes are attack mechanisms—acting as nettles that sting when making contact with other organisms.

There are four major kinds of Cnidaria. Three may be familiar: the jellyfish (Scyphozoa), the hydras (Hydrozoa), and the anemones and corals (Anthozoa). But if you are familiar with the fourth you are probably either a marine biologist or dead. This fourth major kind of cnidarian is the

Cubozoa, or box jellies (also called sea wasps), which are boxy-looking creatures that have the most potent venoms known among animals. All of these organisms share a unique developmental program and can exist in two major body types: polyps and medusae. The polyp morphology is highly characteristic of anthozoans, which do not develop a medusa stage. Polyps can form colonies as they attach to the sea floor, at which point they take up a rather sessile (fixed) existence. The more lively medusa form is characteristic of schyphozoans, hydrozoans, and cubozoans (although all three groups start out as polyps). The difference between polyps and medusae involves a radical reorientation of the mouth. In the polyp form the mouth points up, whereas in the medusa it points down. The rest of the body plans are pretty similar, and it is easy to visualize how a polyp turns into a medusa.

All Cnidaria lack brains; of the dozen or so cell types in the tissues of these animals, however, nerve cells are predominant. These nerve cells form an interesting structure called a neural net, which is an easily observed aspect of these animals and can sometimes be seen with the naked eye. It runs the length of each individual and looks a little like a lattice or a cobweb. It is made up of many nerve cells that make connections to one another. The connections closely resemble synapses and have indeed been called synapses. But are they really the same thing as our synapses? In order for a neural net to be seen as a precursor form to our nervous system, our synapses and theirs should be derived from a common ancestor. Are they?

To look more closely at this question, we need to ask just what a synapse is. Is it simply the "air kissing" of two cells between which chemical or electrical signals course? Or is it more complicated? Let's examine this interesting question by using the rules for calling something the "same" in two different organisms that we established in chapter 1. If we can show that the components of the synapses of the animals we have discussed so far are derived for the different groupings of organisms, then we can answer this question for those groupings. But there are several levels of comparison we can use. There is, first, the structural level, which would include the visible anatomical structure of the synapses concerned. Then there is the genomic level, which involves the kinds of genes implicated in synapse formation. And last, there is the "emergent" level of how the gene products actually interact in the formation of synapses.

In a review of the ultrastructure of synapses in the Cnidaria, Jane Westfall of Kansas State University examined all four major kinds of cnidarians and concluded that cnidarian synapses have a reasonable degree of microscopic similarity to the synapses of vertebrates such as ourselves. But she also pointed out that there are two major ultrastructural differences between cnidarian synapses and those of other organisms. The first is seen in the junctions of muscles and nerve cells (neuromuscular junctions). In higher animals, there are folds at the junctions of these cells that appear after the junction occurs. In Cnidaria, there are no folds. The second concerns the vesicles that harbor chemicals like neurotransmitters (see chapter 2). Cnidaria have many fewer vesicles, and when they do have them they are much larger than the vesicles in our synapses. This suggests that the actual synaptic function of cnidarian vesicles might be different than in higher animals. So ultrastructure is somewhat equivocal when it comes to the degree of "sameness" between cnidarian synapses and those of other organisms. But if we look at the genes present in the lower organisms examined so far, we see something interesting. Both fungi and choanoflagellates have some of the basic signal transduction functions of multicellular organisms, but they lack the genes for making synapses. Still, when we compare sponge, placozoan, and cnidarian genomes we find that sponges do indeed have some of the genes that are needed to make synapses. They are said to have a "presynaptic scaffold" of genes. And remarkably, Placozoa have nearly the same complement of genes for making synapses that Cnidaria have. The plot thickens, and we will return to this point.

Brainness?

Whether or not these primitive animals have synapses, it is still clear that they don't have what we could reasonably describe as brains. So when did the first brain evolve? To delve into this problem we need to know the players in the "higher animal" game a bit better. So far, we have looked at Porifera, Cnidaria, and Placozoa, which are rather asymmetrical creatures. If we split them down what might be the middle of their bodies, the two halves would not look alike or be mirror images. So we need to turn to the bilaterians, so called because their halves would look symmetrical if split down the middle. All Bilateria (except for a small group called Priapulida) also have what is called a coelom, a cavity in the body plan formed as a result of splitting of the mesoderm layer in the embryo. Within Bilateria, the next big division comes as a result of a strange pattern of cell division in the very early stages of development.

After fertilization, the embryo begins a journey of cell division that is controlled by an intricate set of genetic interactions. The first and second divisions into two cells and four cells, respectively, are much the same in all animals. But the next division, to eight cells, starts to differentiate the two major kinds of bilateral animals. In one of these large branches, the third cleavage of the cells is symmetrical (radial cleavage). But in the other major branch of bilaterians, the cleavage causes the cells to spiral across the early embryo. The radially dividing lineage is called Deuterostomia (or deuterostomes), and the spirally dividing lineage is called Protostomia (protostomes). Deuterostomia includes all of the vertebrates, including humans, plus some strange organisms with "false" or partial vertebral columns. Oddly enough, it also includes organisms such as starfish and sea urchins. In contrast, a fruit fly or a nematode has the protostome spiral version of cleavage in the early embryonic stage. The spiral and radial cleavage routes produce an odd situation when the body layouts of protostomes and deuterostomes are compared. Protostomes have anuses where deuterostomes have mouths, and vice versa.

This major division is important when we consider whether a nematode or a fly brain is really a brain like ours. Any time there is a major division of organisms such as the protostome-deuterostome split, we see the results of lineage divergence from a common ancestor. These big divisions are the bread and butter of evolutionary biology, and reconstructing these hypothetical common ancestors often provides clues to the evolutionary pathways involved. But what the common ancestor of the bilaterians looked like is still somewhat mysterious. Because there are as yet no fossils known along this lineage, there is little out there to help scientists interpret the transitions that might have happened along the lineage from the

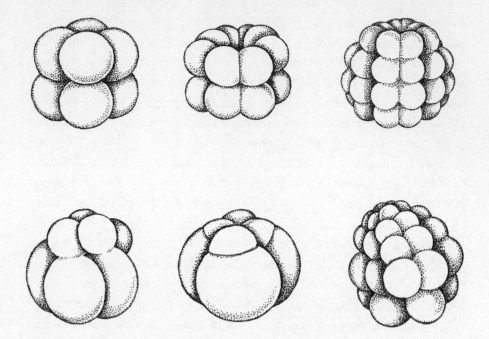

Spiral (bottom) and radial (top) cleavage. The cells in a developing organism are shown for the first several divisions of the embryo. Spiral cleavage is characteristic of protostomes, and radial cleavage is characteristic of deuterostomes.

common ancestor of bilaterians to the deuterostome-protostome split. It is assumed that the genomic and anatomical complexity of animals increased from the common ancestor of the Bilateria and the lower animals like Cnidaria, Porifera, and Placozoa. But because of the lack of fossils, many interpretations of what happened are possible. As the University of California at San Diego biologist Nicholas Holland suggests: "For some, it [the ancestor] is an unprepossessing little creature with no coelom, whereas for others, it is a relatively large coelomate animal." But he adds that "in spite of these uncertainties . . . there is broad agreement that it had a Central Nervous System." This means that the ancestor most likely had peripheral nerves and a diffuse point where the nerve cells communicated.

The upshot is that the examination of what a fly or nematode brain really is can be reformulated in many ways. One way is ask, "What is a head?" After all, the brain resides in the head, right? Another way to think about it is to ask, "Does a fly or a worm brain have the same parts that a vertebrate brain has?" If, for instance, the vertebrate brain has three major parts to it, and we can show that the putative brain of a fly or a nematode has the same parts, then this is a significant step toward understanding its "brainness." Yet another way to ask this question would be to ask, "Are the same genes involved in making the brain of an invertebrate (nonvertebrate: not a wonderful definition for a group, but it works here) and a vertebrate?" Yet another approach would be to ask, "Are the cognitive functions of an invertebrate brain the same as a vertebrate brain?" And again, the real challenge lies in determining whether the common ancestor of these organisms had the same characters. We really need to ask the three initial questions first, in pursuit of the answer to the last.

What are the shapes of invertebrate brains, and where are they? Let's start with one of the simplest invertebrate brains, that of the nematode. As we've mentioned, the entire tiny, wormlike female nematode is made up of a little over 1,000 cells, of which 302 are differentiated neural cells. A clump of neural cells forming a ring at the nematode's anterior end is what most scientists consider the nematode's "brain." Sensory organs (see chapter 4) send signals to this ringlike structure. The ring is connected in turn to a nerve cord that runs the length of the animal on its stomach side. This ventral nerve cord is then connected to muscles. But while the nematode nervous system is a pretty basic structure, these tiny creatures can have pretty elaborate behaviors. For instance, some nematodes can sense and "count" the number of other nematodes near them. If the nematode perceives only a few neighbors, the nematode stays; if it perceives that a party with lots of other nematodes is in progress, it skedaddles. Many invertebrate species have this ringlike brain, including the echinoderms: starfish and sea urchins. (Remember: the echinoderms are more closely related to us than they are to arthropods, worms, and other invertebrates.) We will return to brain shape later in this chapter.

Since the last century, scientists have been using a coloring-book technique to figure out where certain genes are expressed in tissues. One of the tried-and-true approaches is to take an antibody to a protein that is important in development, label it with a florescent dye, and get it to react with a developing embryo. Because antibodies work by sticking to their target, if we make an antibody that is specific for a particular protein, the antibody will adhere to that target protein. In our coloring-book metaphor, the embryo acts like the coloring book, and the positions of the proteins in the embryo are the outlines of the drawings. Because it sticks to the specific protein for which it is made, the antibody acts to position the crayon, and the crayon itself is the fluorescent label on the antibody. When we compile the sequence, we get beautiful pictures of developing embryos, or their organs, colored in where certain proteins are produced. This approach has been used extensively in modern developmental biology, and it is a critical technique when examining the expression patterns of proteins that are important in the brain.

Often what researchers call a brain is no more than a clump of neural cells. Some researchers, though, claim that the brains of invertebrates have the same three-tiered structure that vertebrate brains have. This tripartite structure includes a hindbrain, midbrain, and forebrain. But the evidence for the tripartite brain in invertebrates concerns the expression of genes important in the development of the midbrain-hindbrain junction, rather than the shape of the resulting structure. Vertebrates, and some invertebrates, express the same genes in the brain, at what some believe is a highly conserved midbrain-forebrain junction. But of course these genes might just have been recruited from some more ancient function.

Relevant to all this is an interesting case of "sameness" in which we can ask a peculiar question: "When is a particular kind of organism not that kind of organism?" This conundrum concerns worms. The parasite expert P. Z. Meyers at the University of Minnesota says: "Take the worm. We take the generic worm for granted in biology: it's a bilaterally symmetric muscular tube with a hydrostatic skeleton which propels itself through a medium with sinuous undulations, and with most of its sense organs concentrated in the forward end." If a worm is as Meyers defines, then nematodes are worms. But, perhaps rather petulantly, Meyers realizes that things might not actually be this simple when he says "Sometimes, I confess, this whole common descent thing gets in the way and is really annoying." What Meyers is referring to is the fact that not all worms so defined can be traced back to a single common ancestor to the exclusion of non-wormy creatures. Meyers is annoyed, because he would like to think that a worm is a worm is a worm.

But here is why a worm is a worm is NOT a worm (with apologies to Gertrude Stein). Nematodes and acoelomate worms are certainly both bilateral animals, as are a number of other wormy creatures that we also need to add to the fray: organisms such as annelids (leeches), platyhelminthes (like planaria: small flatworms with pointed heads that can be cut up and will regenerate), and marine wormlike animals called Nemertea. The main difference between the acoelomates and the nematodes is that nematodes molt their skins. Now, this simple molting trait is extremely important in keeping track of the wormlike organisms we have discussed because the nematodes, it turns out, are more closely related to other creatures that molt and are called Ecdysozoa (because a hormone called ecdysone is involved in the molting) than they are to all the other nonmolting wormlike organisms. These are placed in a group of strange animals called Lophotrochozoa (because most animals in this group have a feeding apparatus called a lophophore). The Lophotrochozoa are now classified in a huge group of animals that includes mollusks (oysters, clams, squids) and bryozoans (barnacles).

The systematist's lot is not always a happy or an easy one, and not infrequently scientists just dump a lot of hard-to-classify animals into a convenient group simply to get rid of the problem: out of sight, out of mind. But when their genes are examined for clues to their relationships, this particular garbage pail of organisms turns out to make sense even though they are incredibly diverse and may look hugely different from one another. Our colleagues Ward Wheeler at the American Museum of Natural History and Gonzalo Giribet at Harvard University have examined the largest assemblage of Lophotrochozoa to date and can clearly show common ancestry for all of these bizarre animals. So here is a case where some worms are acoelomate worms and others are molting worms, with the latter more closely related to other things that molt, such as insects and other arthropods.

This discovery in turn has implications for their sensory systems, since it means that the "eye spots" we see in the Lophotrochozoa are probably not the same kind of eye spots we see in nematodes. And in this context of light-sensing cells or the lack thereof, two last lophotrochozoans deserve mention. One is from a crazy group of organisms called the Mesozoa. Incredibly, these animals have only ten to at most forty cells each, with only a few cell types. So why don't we consider this the most primitive animal on the planet, instead of relying on sponges and perhaps Placozoa as the most primitive? Well, it seems that the Mesozoa got to their forty-cell state by "devolving," or losing their cells, so their ancestor was more complex than they are today. Now, with at most forty cells in the entire body plan, there is no room for specialized nerve cells. In fact, these animals (there are about a hundred species of them) have only two cell layers. They are parasites living in the kidneys of marine organisms such as the bivalves and octopus, and basically they absorb nutrients directly through their cells. Mesozoa are considered Platyhelminthes by most biologists, and other Platyhelminthes have eye spots and light-sensing organs, as already mentioned. These communicate with rather simple nervous systems and probably do nothing more than elicit a simple response to light.

The other famous worm is *Buddenbrockia*. Despite a crazy shape and lifestyle, this worm is a worm IS a worm. It is a highly derived animal, even though most of its close relatives are merely single-celled organisms. These single-celled animals are called Myxozoa (don't confuse these with Mesozoa), and the whole group is also embedded within the Platyhelminthes. So single-celledness evidently once more represents a "devolution" of the ancestral wormlike form.

Well, what about the head? "No head, no brain" seems like a reasonable proposition. But is the converse—if you have a head, you have a brain—also true? Because the head is a three-dimensional structure, we can look at this problem from the front end of the body to its back end (anterior to posterior axis) and from the backside of the organism to its belly side (dorsal to ventral axis). It doesn't matter in which dimension we do this: we get the same result when it comes to the head. All invertebrate protostomes have recognizable heads, and it even turns out that the same genes that are involved in making invertebrate heads are in general also involved in making vertebrate heads. In the anterior-posterior axis, the collinear expression of a group of genes important in segmentation and segment-identity in fruit flies (the homeotic genes: HomC) was found to be nearly identical to the collinear expression of the same genes in vertebrates. In fact, the pattern of similarity extends along the entire length of the central nervous system in both kinds of organisms. In addition, genes that are expressed in the most anterior parts of the invertebrate head are also responsible for development in the forebrain of vertebrates. One in particular is a gene called "empty spiracles" (ems), which controls the development of spiracles in fly larvae and is involved in fly brain development. The protein product of this gene, the EMX protein, is important in the development of the olfactory (smell) clusters of neurons called glomeruli in both mice and flies, suggesting that there is a correspondence between the olfactory centers of vertebrates and insects.

Another trick that scientists use to determine the equivalency of genes from, say, flies and mice, involves a technique called "transphyletic rescue." In such experiments, a gene from one species is transposed into another species that lacks the gene. If the gene is important enough in the development of the organism and it's missing altogether, the individuals without it will die at an early stage of development. So when the gene from another species is transposed into the species lacking the gene and the offspring are okay, then a transphyletic rescue has occurred and the equivalency of the specific gene products from the donor species and recipient species are confirmed. Such studies have been done for a large number of genes involved in the head, showing that transphyletic rescue seems to be the rule and confirming the assumption that protein functions are as a rule equivalent across large phylogenetic distances. So far so good, as far as the anterior-posterior axis is concerned. What about the dorsal-ventral axis?

Getting Up Close and Personal with the Fruit Fly Brain

On an even finer scale, in a Herculean study scientists in Taiwan tagged more than sixteen thousand of the fruit fly brain's hundred thousand cells to visualize the connections made among them. Ann-Shyn Chiang and colleagues genetically engineered females of the common lab fruit fly, *Drosophila melanogaster*, with several genes that are active in the growth of nerve cells. They also used what is called a "flip-out construct," which turns everything on at the start of cell division, because it uses a "flipase" gene that is important in the initiation of mitosis. To visualize the nerve cells, they connected a green fluorescent protein to the promoter regions of the genes active in nerve cell formation. To get single nerve cells to show up, instead of all the cells at once, they also put a heat-shock switch on the whole genetically engineered construct, so that they could turn the green fluorescent protein on and off at different times during the development of the fly's brain simply by heating the fly up. By using these constructs and heating flies up at different times, they were able to get thousands of single cells to light up with fluorescence. By organizing the flies to coincide with developmental stages and using computer algorithms to standardize the shapes of the thousands of brains they used, the researchers were able to map the positions of the sixteen thousand neurons tagged. The experiments required huge computers to store the images and process the thousands of brains visually, and the end product is an atlas of the fly brain called FlyCircuit (www.flycircuit.tw), where the sixteen thousand cells and their locations are exquisitely mapped.

The shapes and patterns that the sixteen thousand cells make are intrinsically fascinating. But what is really interesting is how the neurons organize themselves. With this large amount of data, the researchers had to redefine important units of the fly brain. One new unit that they defined is called a Local Processing Unit (LPU) that they viewed as "a brain region consisting of its own LN (neurons) population whose nerve fibers are completely restricted to that region. Further, each LPU is contacted by at least one neural tract." They found forty-one of these in each hemisphere of the fly brains they examined. Another novel unit they defined is what they call a "tract." These are clumps of neural cells that connect different parts of the brain. The fly brain has fifty-eight of these. Last, they determined how many "hubs" the fly brain contained. A hub is simply where the LPUs are connected into discrete clusters. There are six per hemisphere of these in the fly brain. When they put all of the LPUs, tracts, and hubs together, they almost completely filled the volume of the fly brain, suggesting that they hadn't missed much by focusing on the sixteen thousand cells they examined out of the hundred thousand. What is even more tantalizing is the suggestion that the wiring and organization of the fly brain follow the same kind of wiring that computer scientists have developed for the mammalian cortex. Specifically, the fly and mammalian brains both conform to the "small world" principle. This is an organizing principle of the brain where a brain has, as journalist Nicholas Wade puts it, "high local clustering of neurons, together with long-range connections." But similar or not, the question of the brainness of invertebrate nerve ganglia still comes down to the phylogenetic question that we have raised several times already. Beware of not recognizing the metaphor!

Upside-Down Drawings

The importance of getting the metaphor right when it comes to the vertebrate and invertebrate dorsal-ventral axis is evident in a debate that is already almost two centuries old. The controversy started when one of the most innovative and bold scientists of the early 1800s, Étienne Geoffroy Saint-Hilaire, published a picture of an upside-down lobster. His point in doing this was to suggest that invertebrates are in essence inverted vertebrates. The lobster nervous system, like that of all protostome invertebrates, is on its belly side, with the digestive tract on the back (dorsal) side. Think about this—our nervous system is on our back (dorsal) side, and our digestive system is on our belly side. So, metaphorically at least, something has been flipped. Geoffroy had little more than intuition to guide him to this conclusion, but in 1996 Eddie De Robertis and Yoshiki Sasai at UCLA used gene product localization approaches to show that two genes called decapentaplegic (dpp) and short gastrulation (sog), both known to be important for dorsal-ventral polarity in flies, had strong sequence similarity to two genes in the African clawed frog, *Xenopus laevis*. The fly gene dpp showed similarity to the frog gene called Bone Morphogen Protein 4 (Bmp4), and the fly gene sog showed similarity to the *Xenopus* gene chordin (chd). Of interest is that examination of the position and specific interactions of these genes in fly embryos and frog embryos uncovered a curious parallel. Specifically, one of the gene products in each of the two kinds of embryos inhibited the other. So sog inhibits dpp in the fly, and chd inhibits

Bmp4 in the frog. Both dpp and Bmp4 produce what are called signaling molecules, which control a cascade of other interactions of proteins. Curiously, dpp is produced in the dorsal side of the fly embryo, and Bmp4 is produced in the ventral side of the frog embryo. Because these signals need to be localized, in flies sog is produced in the ventral area, and in frogs chd is produced in the dorsal area. In other words, the locations of the gene products have inverted themselves. Subsequent examination of these genes in other organisms has borne out the general pattern that dpp-like genes in protostome invertebrates are produced in the dorsal region of the embryo and ddp-like genes (like Bmp4) are produced in the ventral region of vertebrate embryos. Hence, what appeared to be one of the firmest arguments for lack of homology of the vertebrate brain with the invertebrate brain can be explained by the mere inversion of axes in the embryo. Perhaps the brains of these two kinds of organisms are the same after all.

Tracking Brains in "Almost" Vertebrates

Although we have left some of the questions concerning the origin of synapses and brains for the close of this chapter, it is pretty clear that all vertebrate brains are inherited from a single common ancestor—namely, the ancestor of the chordates (the group that contains vertebrates and some close relatives we'll meet in a moment). But we are skipping over a large group of animals that, as we pointed out earlier, have traditionally been included in that clumsy grouping Invertebrata but are, oddly enough, more closely related to us vertebrates than to other invertebrates. These are the Echinodermata (echinoderms) and the Hemichordata (hemichordates). We have already discussed the Echinodermata, or sea urchins and starfishes. These organisms are radially symmetrical, with what is called pentaradial (five-pronged) symmetry. Most echinoderms don't have a cephalic ganglion; in other words, they lack a brain. What they do have is a neural netlike nervous system, and part of it forms a ring around the esophageal (feeding) tube that is reminiscent of the nematode nervous system but is most likely not the same thing. Whether echinoderms ancestrally had and then lost the cephalic ganglion, or never had one, is a central issue in the quest for the origin of our brains.

One factor that complicates the story of echinoderm brains is that one large group of echinoderms, the Holothuroidea (sea cucumbers), do possess cephalic ganglia. Complicating factor though this doubtless is, it may help us understand how scientists make decisions about reconstructing ancestors. In one scenario, the echinoderm ancestor started out with a central ganglion (a brain), lost it in all its descendants, and then regained it in the lineage leading to sea cucumbers. In the other scenario, the ancestor didn't have a brain, but gained one in the sea cucumbers. The second scenario takes fewer steps than the other—two steps versus three—and, other things being equal, we would incline toward the shorter route. Evolution, however, does not necessarily take the shortest path. And we haven't yet said anything about how the ancestor of echinoderms was reconstructed. How the ancestor started off is critical to knowing how the brain evolved in the rest of this large group of interesting animals.

The other group of nonchordate deuterostomes—Hemichordata—is made up of three major kinds of wormlike animals: the Pterobranchia (pterobranchs), the Enteropneusta (acorn worms), and a single, hard-to-place species that forms the Planctosphaeroidea. These organisms have three body segments—protosome, mesosome, and metasome (from mouth end of the body to the butt end)—and a weblike nervous system that runs the length of the body, although there is no brain to speak of. But genes that are responsible for the positioning of the nervous system in chordates *are* present in hemichordates and are expressed in much the same pattern. To see this, we need to go back to the three-part body structure of hemichordates. Genes expressed in the chordate forebrain and midbrain are expressed in the protosome of the hemichordate, and genes expressed in the chordate hindbrain are expressed in the mesosome of hemichordates.

Although this superficial similarity of expression patterns is intriguing, some researchers have suggested that the ancestral state of the deuterostomes actually was a diffuse neural net. In fact, examination of the hemichordate nervous system shows that this diffuse net lies in the epidermis (the basiepidermis, to be exact). The position of this diffuse nerve net has prompted Nicholas Holland to suggest that the ancestor of chordates had a "skin brain" and that the more localized nervous systems of chordates evolved after hemichordates and echinoderms (which also have the diffuse skin brain) diverged from chordates. A diffuse state of the ancestral deuterostome nervous system would also aid in interpreting the brain transitions in the echinoderms. If the ancestor of the deuterostomes had a diffuse neural net without a brain, then the ancestor of echinoderms would most likely have lacked a brain, and the second echinoderm brain scenario we described above would become the preferred interpretation.

If this diffuse neural net idea is correct, as chordates evolved, the diffuse nervous system of the common ancestor of chordates coalesced into more localized and well-defined structures. In this ancestor, a notochord (a rodlike skeletal structure that runs nearly the length of the animal) formed on the dorsal side of the animal above the gut. The notochord is not the backbone, nor is it the vertebral column or the spinal or nerve cord. Instead, it is a rod of tissue that shows up early in embryological development and more or less organizes the subsequent development of structures near it. As embryogenesis proceeds, the nerve cord develops on the dorsal side of (above) the notochord. Further along in embryogenesis, the nerve cord differentiates at its anterior end into the brain and posterior to the head into the spinal cord, while the notochord itself starts to disappear.

The first organisms to diverge from the chordate ancestor were the Urochordata, or tunicates. These are extremely simple animals that have been described as no more than a sac through which water is filtered. And the nervous systems of adult tunicates are so primitive as to be misleading. Only a single small ganglion of neural cells resides in this filtering sack, with minimal nerves sent out to the organs. But these little guys are actually much more complex than that, because as babies (larvae), they have a full nervous system that would do any chordate proud, with a well-developed dorsal nerve cord and sense organs at the anterior (head) end. Only after they have metamorphosed into adults do their nervous systems take on their ultra-simple aspect.

The next group of organisms to diverge from the remaining chordates is Cephalochordata. These are little marine creatures also known as amphioxus or lancelets. They are a bit more complex than the Urochordata, and their nervous systems are only a slight "improvement" over adult tunicates. A nerve cord runs through the length of the animal, and at the anterior end is a clump of cells at best barely recognizable as anything special. In fact, many zoologists doubt the presence of any sort of "brain" in these simple organisms. To sum this all up, for the chordates, brains come and go in some scenarios and go and come in others. One thing for certain, though, is that once brains came into focus for vertebrates, they stayed.

Cyclostomes

It is almost certain that the common ancestor of all today's vertebrates (fishes, amphibians, reptiles, and mammals) had a brain. After all, an alternative favored by many is to call them Craniata, or things with crania (a cranium is a skull without the lower jaw). And if you have a cranium, you have a brain. But there is an interesting question about the relationships of early vertebrates with skulls that we need to look at before considering the organization of vertebrate brains (chapter 4). In this general part of the Tree of Life there are two rather intriguing groups of animals with vertebrae that have bothered systematists for a long time. One is the hagfish, which form a group called Hyperotreti that lacks a skull altogether. The other is the lampreys (Hyperoartia), which have a cranium but no jaw. It was long thought that the lampreys were more closely related to the rest of vertebrates than to hagfish, because lampreys have crania and hagfish do not. But work published in 2010 by Alysha Heimberg at Dartmouth and her collaborators on small RNA molecules has suggested a much closer relationship between lampreys and hagfish, which together form a group called Cyclostomatidia (cyclostomes).

So in the classical scenario, where the lamprey is more closely related to other vertebrates than to the hagfish (and cyclostomes are not monophyletic), the ancestor of vertebrates is reconstructed as an organism with a simple body plan, without a skull, and with a rudimentary nervous system and brain. But, as Heimberg and colleagues point out, "with the recognition of cyclostome monophyly . . . that evolutionary insight is lost. Evidently, the ancestor of vertebrates was more complex, phenotypically and developmentally, than has been perceived hitherto." What this means is that the ancestor of all vertebrates carried the potential to be much more

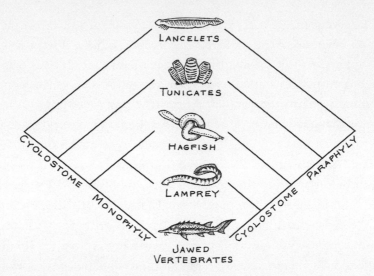

The difference in the simple rearrangement of a critical pair of taxa in the vertebrate Tree of Life. Cyclostomes (hagfish and lamprey) are monophyletic in the tree on the left and paraphyletic in the tree on the right. Heimberg and colleagues suggest that cyclostome monophyly is the best explanation for the genealogy of the vertebrates.

advanced than the so-called primitive vertebrates—like the jawless lamprey and hagfish—that we see today on the planet. As we have stressed, the choice of outgroup is extremely important in interpreting the evolutionary events leading up to the vertebrate brain. In this case, the classical outgroup to the vertebrates is the hagfish, and a simple interpretation of how the brain evolved can be made. But if the outgroup is in fact the entire cyclostome group, a new interpretation is required.

When Appearances Can Be Deceiving

It is now time to take stock of what our ramble through this part of the Tree of Life has told us. We started the chapter by looking at how cells communicate with each other, and we began to ask questions about what implements communication between cells. This exploration led us to an examination of what a nervous system is and isn't, and also to inquire what a synapse is and isn't. We found that although plants don't have synapses, they do have structures that are metaphorically similar to animal synapses. We also found that some of the more primitive animals, such as placozoans and sponges, don't have nerve cells or synapses and hence have no nervous systems. But we did find that these nerveless animals have some of the genes necessary for making synapses. The simplest animals with actual synapses are the cnidarians, but instead of anything you could call a brain, these animals have a neural net. The classical view of the synapse is that neural cells of all animals are the same in origin and that the synapses of invertebrates and vertebrates are thus the same. In this old view, what makes animal nervous systems different is not the kind of cells present but the number of neural cells that animals possess.

What's more, when we looked more closely at the synapses of animals with nerve cells, we found two important things. First, organisms without nerve cells actually have genes that make proteins that are found in synapses. Fungi have genes involved in membrane activities that are precisely the same genes higher organisms use in making synapses. Fungi just use them for different functions, such as regulating protein synthesis and altering cell structure. Organisms, such as sponges and Placozoa, that lack synapses and nerve cells nonetheless have nearly full complements of genes used elsewhere to make nerve cells and synapses. Second, we found that although invertebrates and vertebrates both have synapses, vertebrates have many more genes for synapses than invertebrates do. Specifically, vertebrates have expanded the number of genes that code for proteins that make up the synapse, among them receptors, adhesion proteins, and proteins involved in membrane scaffolding. When the kinds of proteins that make up the synapse of an invertebrate such as the fruit fly are compared to the proteins that make up the mouse synapse, it is found that the mouse proteins have greater signaling complexity than those of the fly. These results suggest that there is a core, or scaffold, of genes that existed in the common ancestor of all animals that produce synapses. As animal nervous systems have become more and more complex, this complexity has been achieved through the expansion of the number and kinds of genes that make up the synapse. As Richard Emes of Keele University and his colleagues put it, "The evolution of synapse complexity around a core proto-synapse has contributed to invertebrate-vertebrate differences and to brain specialization." It's not the number of neural cells that counts but rather what's in the neural cells and, more specifically, what's at the synapse.

In this context, it might be simplest for us to consider the neural net of cnidarians as something different from the nervous systems of bilaterians. This view is even more attractive when we look at the current understanding of the phylogeny of nonbilaterian animals in relation to Bilateria. During our discussion we have been deliberately vague, because the phylogenetic relationships of these animals are still not that clear, even though many scientists would like to think they are. But basically, when we consider the major groups of organisms involved in this part of the Tree of Life, we are talking about five taxa plus an outgroup: Placozoa (P), Porifera (S), Cnidaria (Cn), Ctenophora (Ct), and Bilateria (B). The number of trees that can be constructed using these five taxa is 105. One of these is shown in the adjacent diagram. This is the classical arrangement, with sponges as the most primitive animals, followed by Placozoa, and then the Cnidaria-Ctenophora pair that is most closely related to Bilateria. It turns out that 5 others of the potential 105 ways of arranging these taxa have also appeared in the literature. Only one of the 6 separates the cnidarians and ctenophores. So if we simply call the Cnidaria-Ctenophora couple a single taxon, then the number of possible trees drops to 15, of which 5 have been offered as solutions for the four taxa. So fully a third of all the possible permutations of the four taxa P, S, Cn, and B have been suggested as possible solutions.

Many scientists prefer the tree topology whereby the sponges are most primitive, followed by Placozoa, followed by the Bilateria-Cnidaria pair. This topology fits well with the idea that synapses, nervous systems, and brains all evolved just once on the tree shown. Any other arrangement would mean that nervous systems and synapses evolved more than once. So this particular tree is certainly pleasing if synapses and brains indeed evolved only once. But could other scenarios be more accurate? What happens if synapses are in fact metaphors? Indeed, what happens if nervous systems are metaphors too? We find that one of the alternative trees offers a very interesting scenario. This is the one suggesting that Bilateria were the first group to diverge from the common ancestor of all animals. Next, Placozoa

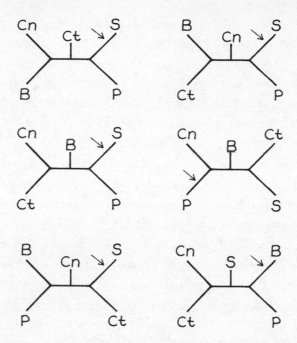

Alternative trees for arranging Sponges (S), Cnidaria (Cn), Ctenophores (Ct), Placozoa (P), and Bilateria (B). The choanoflagellate *Monosiga* is used as an outgroup, and the arrow indicates where *Monosiga* would root each of the six hypotheses. With a little imagination, one can see the subtle differences in the six hypotheses. For instance, the top two trees would have Porifera (sponges) as the most "ancient" metazoan but switch the position of Ctenophora and Cnidaria. The bottom right tree is interesting in suggesting that Bilateria is the most "ancient" metazoan lineage.

branched off, followed by the sponge-Cnidaria pair. In this scenario, synapselike structures evolved twice, but this isn't so unreasonable given that we have already established that the cnidarian and bilaterian synapses have basic complexity differences—and, more important, that cnidarians didn't have to evolve new genes to make synapses. And if synapse genes are present in Placozoa and sponges, then they were also present in the common ancestor of cnidarians. This scenario also implies that the neural net of Cnidaria is different from the nervous system of Bilateria, which prompts us to suggest calling the cnidarian neural net a "cneural net." It isn't the same as a bilaterian nervous system, so why spell it the same way?

Through looking at bilaterian nervous systems, we started to get at the question "What is a brain?" We examined this by looking at some of our closer relatives, the bilateral invertebrates, and we got a pretty good idea of what scientists have had to struggle with in approaching this question. In this context, it is maybe helpful to look at popular definitions of the word "brain," which seem to come in two basic kinds. First, there is what we might call the "vertebrocentric" definition, a definition that focuses on what vertebrates have and ignores invertebrates completely. The second is the "obligatory invert-mention" definition. Such definitions come in two parts. The first part is a detailed definition of the vertebrate brain, followed by the obligatory mention that invertebrates have something that we can call a brain. This second kind of definition, the one involving invertebrates, is interesting largely because it opens a can of worms (so to speak). Some of these invert-obligatory definitions suggest vaguely that the invertebrate brain is "functionally similar," or is a "comparable organ" to that of verte- brates. Other definitions of the invertebrate brain are a little more precise, suggesting that it is "comparable in position and function to the vertebrate brain"; in effect, though, this definition is nothing more than a combina- tion of "comparable organ" and "functionally similar."

Definitions of the brain in scientific texts are more precise but not much better. One we came across stated that the invertebrate brain is "a part of the nervous system *more or less* corresponding to the brain of vertebrates" (italics ours). Let's look at this statement in detail, because its imprecision is at the heart of the problems involved in stating that structures are "the same" from organism to organism. In chapter 1 we introduced an important concept with respect to interpreting patterns of evolution among organisms. For the lack of a better word, we called it "sameness." But there is a more technical term for this idea, namely "homology." The nineteenth-century British anat- omist Richard Owen, who coined the word "dinosaur," coined this one, too. His definition of homology was important, because it helped scientists avoid the "comparing apples and oranges" problem that they faced when associat- ing things. That definition, "the same organ in different animals under every variety of form and function," was proposed in the 1840s, well before Darwin and Wallace presented their ideas about natural selection and descent with modification. So when Darwin and Wallace presented these radical notions about the nature of organismal change, the definition had to be altered to include descent with modification. In chapter 1 we defined three major kinds of changes that can occur when we place things in the context of descent with modification: shared and derived changes, primitive changes, and convergences. Of these three kinds of changes, only the shared and derived changes are homologies. So the best way for us to examine whether something in one organism is the "same" in another is to make sure that it is shared and derived—that is, inherited from the same common ancestor, and neither a convergent change nor a primitive feature.

Now let's return to the "more or less" part of the definition. In the 1970s, molecular biologists started to sequence genes from a wide variety of organisms. To make sure that the comparisons they were making were appropriate, they needed to establish homology among the genes. Arguments soon broke out, because some scientists were claiming that some genes were, say, 8 percent homologous to others. They based statements like these on how much of the gene sequence matched, or was similar, from one species to another. One scientist in particular, University of California at Davis biologist Walter Fitch, was concerned about this problem, because he knew that similarity was not the proper criterion for establishing homology. In an impish moment of his stellar career, he pointed out that homology is a lot like pregnancy. One cannot be 85 percent pregnant. One is either pregnant or not. Homology is the same. Things are either homologous or not; they cannot be anything other than 100 percent homologous or 0 percent homologous. Hence, just because two things show some similarity, they aren't necessarily homologous.

With these rules in hand, you might think it should be easy to examine the sameness of things. But not so, because there are many ways of dissecting the traits we are interested in. Think of the wings we considered in chapter 1. We discussed this problem by examining the anatomical structure of wings. It was pretty straightforward to figure out the lack of "sameness" of insect wings, bird wings, and bat wings simply by looking at their makeup. The lack of the similar elements in these three different kinds of wings was the lynchpin in determining their homology. But what if we were to examine wings using a genetic approach and found out that there are several common genes involved in producing the wings of these three different kinds of organisms? Would that make a difference? Many scientists have seen this pattern of disparate anatomy but not-so-disparate genetics and have made extravagant claims about homology. This is much like the assertions of the plant neurobiologists we talked about earlier in this chapter: it's all about metaphor. Luckily, metaphors are not like jokes. Any joke you need to explain is a bad one. As E. B. White once said, "Explaining a joke is like dissecting a frog. You understand it better, but the frog dies in the process." On the other hand, any metaphor you need to explain is quite likely a good one. As we suggested earlier in the context of plant nervous systems, metaphors help us explain things better; and understanding the brain in this context of metaphor is actually very useful. Establishing the appropriate metaphors makes the job of the evolutionary biologist much easier.

What all of this means in practice is that, when push comes to shove, it is very possible that our vertebrate brain is just one kind of brain out of many that have evolved during the expansion and divergence of the branches of the Tree of Life. After all, scientists have accepted for some time that eyes have evolved independently more than twenty times in animals. If eyes can do it, then perhaps brains can, too. If we are going to call our brain THE brain, then yes, vertebrate brains are the only brains that have evolved on this planet. But if we prefer to use verbiage less narrowly, then other brains might have evolved independently several times.

4 MAKING SENSE OF SENSES

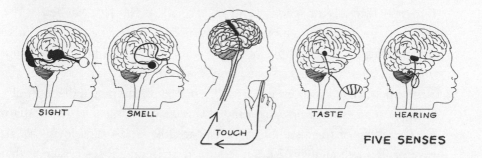

SIGHT SMELL TOUCH TASTE HEARING

FIVE SENSES

From time immemorial people have wondered how other organisms perceive and interpret the world. And as is often the case, we find ourselves referring to Aristotle to get a glimpse of how the ancients viewed the living world around them. Aristotle wrote two treatises on organisms and their place in nature: *On the Parts of Animals* and *The History of Animals*. In these astonishing works, Aristotle posed nearly every question in biology that natural historians would ponder over the next two thousand years. Among other issues, he was concerned about the senses, where these emanated from, and how they worked. He asked some very important questions and got some answers right but others wrong. The following quotation from *The History of Animals* details Aristotle's belief that sensation was controlled partially by the brain but mostly by the heart: "The brain in all animals that have one is laced in the front part of the head, because the direction in which sensation acts is in front; and because the heart from which sensation proceeds, is in the front part of the body; and lastly because the instruments of sensation are the blood-containing parts, and the cavity in the posterior part of the skull is destitute of blood vessels." And although his notion that the senses are processed in the heart is wrong, Aristotle was nonetheless influential in describing what the senses are. He developed a system for them, shown in the adjacent figure from Nils Damann at Katholieke Universiteit Leuven. More than two thousand years later, Aristotle's system can still be mapped onto a notion of how certain gene complexes are involved in specifying the senses. Note that a sixth sense is included in the list; this is not Bruce Willis's *Sixth Sense* but rather the sense of orientation, or balance. Later scholars and artists such as Galileo, Michelangelo, and Leonardo da Vinci (among whom were some superb anatomists) also thought long and hard about how the brain works and how the natural world and all of its sounds, sights, smells, and forms are interpreted by the brain. Michelangelo was even "accused" in 1990 of slipping a diagram of a giant brain into that famous painting of God reaching out a finger to Adam on the ceiling of the Sistine Chapel. Apparently, God's cape forms an almost perfect silhouette of a brain, complete with an angel's posterior that

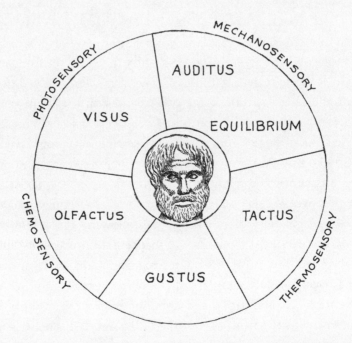

Our six senses, with Aristotle in the center. Auditus = hearing, Visus = vision, Olfactus = smell, Equilibrium = balance, Tactus = touch, and Gustus = taste. Redrawn from Damman et al. 2008.

forms the pons (a part of the brain stem), an angel's knee that forms the pituitary gland, and an angel's scarf that forms the vertebral artery which feeds blood to the brain. Apparently, Michelangelo did a lot of this with his paintings in the Sistine Chapel. In 2010 yet other neuroanatomical drawings were "discovered" in his paintings.

Leonardo's drawings also provide superb examples of the state of knowledge of the brain at this time. But perhaps the most interesting take on the brain and our senses came from Galileo. Rather than adopting an anatomical approach, he developed a unique take on how optics worked while making his observations of the sun and the moon, and he came up with a special way of thinking about the senses—in particular, sight. Galileo felt that nothing really existed until organisms developed the ability to sense phenomena. To him the physical world world—mountains, lava, water, air, everything—simply had no form until organisms that could sense the environment appeared.

Sensory Evolution

As we examine the senses, let's continue to do so from an evolutionary perspective—that is, by viewing the human brain as the product of millions of years of evolution. For that matter, any brain or nervous system we examine—from the so-called neural net of jellyfish to the slightly more complex nervous system of a nematode to the seemingly overly complex nervous systems of vertebrates—is of course the product of millions of years of evolution. To retrace this evolutionary journey through the senses that contribute so much of the input to our nervous systems, several organisms on the Tree of Life are particularly helpful. Every living (and extinct) organism had a common ancestor with us at some point in evolutionary history. For instance, our common ancestor with bacteria lived about 3.5 billion years ago, and our common ancestor with yeast lived about 1 billion years ago. More recent are our common ancestor with insects (550 million years) and that with lizards (about 400 million years). At the more recent end of the scale, we last shared an ancestor with rhesus macaques at around 22 million years ago and with chimpanzees (both common chimps and bonobos) at between 6 and 8 million years ago. And our most recent common ancestors with extinct relatives lie within our own genus *Homo*. Last, we certainly can't forget that all *Homo sapiens* on this planet today are related to each other through a common ancestor who lived not much more than 150,000 years ago.

So, as we examine our senses and consider how they arose, in each case we will need to survey the sensory systems of a succession of other species. In the case of ancient systems, we can use single-celled organisms or very primitive animals such as sponges and cnidarians. For a look at how more complex forms of animal sensing arose, we can use studies on the sensing systems of organisms such as *Drosophila melanogaster* and the nematode *Caenorhabditis elegans*. Because the next important phase in the evolution of our brain involves vertebrate neural systems, we can use primitive vertebrates like fish and lizards and birds to detail the senses at the relevant evolutionary steps. And finally, we will look at how mammalian sensing systems arose and what the components of these systems are. It is relatively easy to examine the senses in this way, because for a century now, scientists have been using creatures like mice, fruit flies, yeast, and bacteria as so-called model organisms. More recently, agricultural animals such as the chicken and developmentally important animals like nematodes, zebra fish, and sea squirts have been used as model organisms, enhancing the range of our knowledge about how senses work.

Let There Be Light: In the Beginning There Were GPCRs

Light is actually a mixture of different forces produced by the interaction of charged particles such as electrons and protons. Because it is a combination of the fields produced from electric charges and magnetic forces, light is a form of electromagnetic radiation. Light can be looked at as both particulate and wavelike, and it is via the wavelike properties of electromagnetic radiation that we perceive the forms and colors of objects in nature.

As with any wave system, different waves can have different properties. Most important for how we perceive electromagnetic radiation is wavelength. Technically, light has a broad range of wavelengths and includes phenomena such as radio waves and microwaves. Radio waves have lengths up to kilometers long, whereas microwaves only about 30 centimeters long implement the heating of food in those kitchen appliances on which many of us rely. At the other end of the spectrum of light are gamma rays and X-rays. Gamma rays have minuscule wavelengths that are more than fourteen orders of magnitude smaller than radio waves. And among this sea of waves, there is only a small range of wavelengths that we detect as light. These vary from the edge of the infrared range to the edge of the ultraviolet range—namely, from 0.7 micrometers to 0.4 micrometers. A micrometer is one billionth of a meter, so this means that our eyes and brain not only can detect tiny wavelengths of light but also can tell the difference between wavelengths to an amazing accuracy of less than a billionth of a meter.

How does our visual system do this? The behavior of molecules provides the simplest answer for how the information gets as far as the brain. But that's just the start: how it is translated into what we actually "see" is a bit more complex, and we'll address this later (chapter 5). The external light is collected by our eyes—one of those organs that evolution has tossed around, experimented with, and reinvented several times over the past billion years. The zoologist Ernst Mayr once estimated that eyes have evolved independently in animals more than twenty times.

Scientists now know that many of the photoreceptive proteins in our eyes existed in the common ancestor of all animals and that a specific master switch gene exists in all organisms with eyes (in *Drosophila*, this gene is oddly called "eyeless"), suggesting that eyes are a primitive, homologous trait for animals. If, however, we use our phylogenetic metaphor principle to determine what is and is not an eye, it becomes clear that the many of the different "eyes" in animals are analogies rather than homologies, meaning of course that the eyes of an insect may not be the "same structure" as the eyes of a vertebrate. And indeed they are structurally quite different, with varied cell configurations and genes involved.

For instance, the insect eye is made of many separate facets, or "ommatidia," that are grouped into what is known as a compound eye. Some insects have thousands of ommatidia, others have many fewer. The cells that make up these compound eyes are called rhabdomeres, and they use what are known as rhabdomeric opsins, a category of the important type of

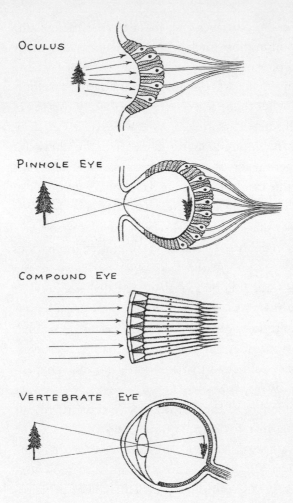

OCULUS

PINHOLE EYE

COMPOUND EYE

VERTEBRATE EYE

Eyes. These drawings represent four of the twenty-five or so kinds of eyes that have evolved on earth. The top diagram is of an oculus, a primitive light-sensing organ found in many kinds of organisms. Below that is a pinhole eye, found in several animals, including the molluscan *Nautilus* and planarians. The eye diagrammed third from the top is a compound eye. This eye is made up of smaller light-sensing subunits called ommatidia. The arrows indicate light coming into the compound eye, hitting the ommatidia. Insects are a good example of organisms with compound eyes. The bottom diagram is of a vertebrate eye with a lens and cornea (to the left of the eye) and retina (the dark patch of tissue to the right of the diagram).

receptor known as opsins. The rhabdomeres are a single cell type that collects and transduces the light signals coming into the compound eyes. Mollusks have very complicated but quite differently formed eyes, and again these eyes are most likely not the same eyes that we see in vertebrates. Once the vertebrate eye evolved, though, vertebrate "eyeness" stuck around pretty tenaciously, so it is difficult to argue that the eyes of all vertebrates are not homologous to one another.

Vertebrate eyes, though not compound, are nonetheless complex. They contain two major kinds of light-sensing cells known as rods and cones.

Both types of cell are long and slender, but they have different light-sensing properties. Depending on what kind of proteins the cell is using to accomplish transduction of the light signal, rods and cones will detect different wavelengths of light. And it is the opsins that are the key to understanding how incoming light is translated into a format the brain can interpret. Opsins make up a complex family of related proteins that all pretty much do the same thing, but each with a different wavelength of light. These molecules are about four hundred amino acids long, and they fold into a beautiful structure similar to that of the G-protein-coupled receptors (GPCRs) we met in chapter 2. They have seven helical regions, with what are called "loops" between the helices. In humans, these proteins sit in the membrane of the specialized rod and cone light receptor cells that lie in the retinas of our eyes. Oddly enough, opsins are not always unique to eyes, for in some other organisms they can also be found in strange places that do not resemble eyes at all. Nonetheless, they also function in these places as light receptors.

Different opsins will detect different wavelengths of light according to the particular amino acid arrangement in their primary (string-of-beads-like) structure. In fact, some amino acids that are located nowhere near the retinal binding pocket of the opsin can have a huge impact on what wavelength of light implements the conversion of the opsin from the "cis" to the "trans" form. These amino acids, known as "spectral tuning residues," implement responses in the cells and are much like the color fine-tuning dials of old TVs. The signal that is thus produced by opsins in the receptor cells now needs to be sent to the brain, and this is accomplished via a G-protein. And if this sounds familiar, that's because, as we noted, opsins have the same structure as the GPCRs introduced in chapter 2.

Opsins have a fascinating evolutionary history. The various proteins in this family have been classified according to both their relatedness and their function. The first division of opsins is into what are called Type I and Type II proteins. Because Type I and Type II opsins have very different sequences at the amino acid level, fit into the membrane in varying ways, and interact with multiple kinds of chromophores, they clearly arose from two very different ancestral proteins. In fact, it isn't too hard to envision that the seven-helix, membrane-spanning structures of Type I and Type II opsins have converged on each other or, as Darwin would have said, are analogies.

Just by themselves, Type II opsins have an interesting evolutionary history. Although the early evolution of Type II opsins is hazy, it is clear that these molecules are related to GPCRs. They have the same three-dimensional structure, are G-coupled, and have similar amino acids in their beadlike structures. And much as we described in chapter 1, the opsin genes expanded into a large and important gene family through gene duplication, and sometimes by wholesale duplication of chromosomes. To date thousands of GPCRs have been sequenced from different organisms: the latest gene tree for GPCRs has more than a hundred thousand genes in it. When the thousands of opsin sequences available are analyzed along with the existing GPCRs, we can ask specific questions, such as "Which came first, the opsin or the GPCR?" Actually this is an easy one, because it is evident that opsins are the latecomers and that GPCRs existed well before they ever specialized into opsins. Next we can ask, "How many Type II opsins are there, and how are they related to each other?" According to the Kyoto University biologists Yoshinori Shichida and Take Matsuyama, there are four major kinds of Type IIs: ciliary opsins, rhabdomeric opsins, photo-isomerases, and neuropsins. These four categories have become diverse because they have evolved to use different G-proteins in their interactions with the cell. The last two categories are somewhat obscure, and little is known yet about them. But this doesn't matter much for us, because the ciliary opsins and one of the rhabdomere opsins are the ones vertebrates have in their visual system. The rhabdomeric opsins are more common in invertebrates like *Drosophila* and mollusks, although one rhabdomeric opsin, melanopsin, is also found in vertebrates and is important in circadian rhythm.

Before we go on to discuss the opsins that give us the ability to see colors, let's examine some interesting cases of how opsins work in actual eyes. *Caenorhabditis elegans*, the tiny nematode, has no opsins in its genome, yet it is able to detect and respond to light, usually by avoiding it. It compensates for not having opsin genes by using a cell receptor that is closely related to the taste receptors that its genome actually does specify. It turns out that nematodes are not alone in this: fruit flies (or at least their maggots) have something similar. The larval stage of a *Drosophila* is caterpillar-like, and its life involves no more than crawling around and feeding. Unlike adult fruit flies, which have complex eyes with hundreds of ommatidia in them, the larvae have eye patches of a mere twelve facets, arranged on each side of the larva in a structure called Bolwig's organ. Researchers in the renowned Jan laboratory, a neurobiology facility at the University of California at San Francisco known for its research on fruit flies, showed that even if they rendered the Bolwig's organs nonfunctional, the resulting larva would still avoid light—and would, in fact, do it as well as a normal larva. But a larva lacking Bolwig's organs responded only to very short-wave light and was unperturbed by red and green spectrum light. It seemed that cells somewhere else in the larva were also detecting light and transmitting information to the larva. And how! It turned out that the larval body of a fruit fly is *covered* in tiny receptors for shortwave light. These receptors are actually very closely related to the taste receptors that *Caenorhabditis elegans* uses to detect light. The moral is that there is more than one way to detect light and, more important, that most organisms will develop such systems if they need them.

Among the light detectors, five kinds of ciliary opsins are by far the most important in our own visual systems and those of other vertebrates. As we've mentioned, the retinas of our eyes, where the incoming light is transformed into neural information, have two kinds of collecting cells: rods and cones. Rods have a single kind of opsin associated with them, while the cones have four different kinds. Rods use the opsin called rhodopsin, which detects monochromatic hues with great sensitivity. The rods are thus important in night vision. The cone cells, on the other hand, are more versatile though less sensitive to light intensity. Because of the four opsins associated with different cone cells, each cell can react individually to four different ranges of wavelengths of light. The wavelengths that cones can detect fall roughly into three categories: long, medium, and short. (If only all biological nomenclatures were this simple!) The long-wavelength opsin, or LWS, reacts to light in the red region of the color spectrum, while the medium-wavelength opsin (MWS) reacts to light in its green region. And the final category actually splits the reaction to short wavelength light between two opsins—SWS1 and SWS2—that detect blue and violet, respectively.

The structural changes in the proteins used to detect different wavelengths of light involve simple variations in primary amino acid sequence. Clearly, all these opsins are somehow related, and one of the questions we can ask is simply, "How?" Interestingly, when the relationships of the five ciliary opsins were established, it became evident that LWS is the oldest sibling of the family, followed by SWS1 and then SWS2. The youngest of the group are the rod cell rhodopsins and medium-wavelength opsin. This rather counterintuitively suggests that in vertebrates the detection of monochromatic light (like many fourth and fifth children in families) was an evolutionary afterthought! The full genomes of a wide array of vertebrates and "near-vertebrates" (those groups of animals that are closely related to vertebrates but lack vertebrae—like sea squirts, acorn worms, and so forth) reveal pretty clearly that the ancestor of all vertebrates had LWS, SWS1, SWS2, and rod cell rhodopsins.

So how did the various combinations of opsins evolve? As we saw in chapter 1, the evolution of gene families is affected not only by divergence of species but also through the duplication of genes. So the ancestral ciliary opsin that existed in the ancestor of vertebrates and near-vertebrates such as echinoderms and tunicates must have duplicated at least twice to produce four basic ciliary opsins in the vertebrate ancestor. The only missing opsin, MWS, arose a hundred million years or so later, via another gene duplication, in the ancestor of vertebrates with jaws. It is amazing that the machinery for pretty good color vision was in place in our ancestors at five hundred million years ago!

But that doesn't mean that all of the organisms illustrated in the adjacent figure had color vision. And it doesn't mean that there is anything special about this five-ciliary opsin gene arrangement. Opsin genes are something of a genomic wild card: they can easily be deleted or duplicated during evolution. This means that various branches of the Tree of Life can have pretty different histories for their opsin genes, yet still have fully functional

Phylogenetic tree for vertebrates, showing where the opsins arose and were eliminated. For instance, the node in the tree where jawless vertebrates diverged from the basal vertebrate lineages of tunicates and lancets at 670 million years ago gained the LWS, SWS1, SWS2, and Rh opsins as a result of a genome duplication. The inset shows the genealogy of the opsin gene family. All of these opsins have a single common ancestor. In this tree, the LWS opsins diverged first. Next the SWS1 opsins diverged, followed by the SWS2 opsins. Finally the Rh opsins diverged from each other. The shades in the diagram represent the wavelengths that the opsins detect. From the bottom, the black shade represents purple to blue wavelength. The next dark gray layer represents blue. The medium gray represents green. The light gray represents green to yellow, and the white area represents red.

color vision. Researchers have accordingly determined that the opsin gene repertoire is extremely complex in some fish groups—for example, the cichlid fishes living in Africa's Lake Victoria. This group of fish exploded in a short period into many species with an astonishing degree of variability in size, shape, and, most important, color. If color is an important component of how the many different species of fish recognize each other, then more precise color-recognizing machinery would be an important evolutionary innovation for them. Indeed, in the cichlid species examined to date, instead of the usual four cone ciliary opsins there are seven, including two extra opsins that detect light in the green wavelength range and one extra opsin that detects light in the violet range.

In contrast, it has been estimated that many of the opsin genes that our vertebrate ancestors possessed have been lost as mammals evolved. It turns out that the complete repertoire of ciliary opsins in the ancestor of vertebrates decreased in the ancestor of the mammals: it went from five to just three. LWS, SWS1, and rhodopsin were all that remained. We are able to infer this from looking at nonplacental mammals (such as the Australian platypus and echidna) and several placental mammals and then reconstructing the ancestor of all mammals accordingly (see chapter 1). The platypus is something of an enigma, because it has apparently retained its SWS2 gene while losing its SWS1 gene. The outcome of all this analysis is that the ancestor of mammals was most likely dichromatic (LWS and SWS1 only). And indeed, most mammals today are dichromatic, with only two kinds of cone cells, one kind with LWS and one with SWS1. Primates, as well as some other scattered groups of mammals, have regained trichromacy (three cone opsins).

This raises the question of how we primates got back to being trichromatic (LWS, SWS, and MWS)—and even, in rare cases, tetrachromatic (LWS, SWS1, SWS2, and MWS). But if these scenarios sound a bit odd, they get even messier when we look more closely. For there are two ways to be a trichromatic primate, and we can best understand what happens if we look first at how primates are related to one another. Over the past fifty million years or so the ancestor of modern primates gave rise to over three hundred living species and a plethora of now-extinct ones. The first extant

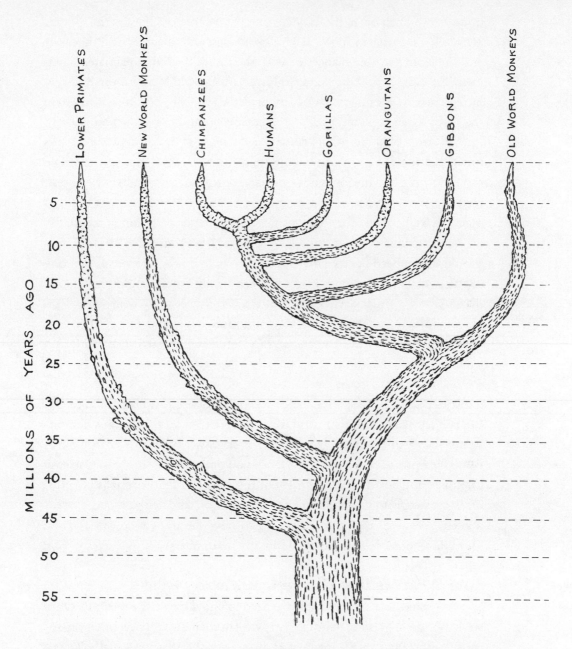

Primate tree showing the branching events in this family of mammals. The timescale for divergence events is shown on the left.

lineage to branch off in the primate part of the Tree of Life was the "lower primate" group that includes the lemurs, lorises, and galagos. Next, the tarsiers split from a common ancestor with the rest of the living primates. Then came the New World monkeys, followed by the Old World monkeys. The living species of our branch of the primate tree, the hominoids, include the gibbons, orangutans, gorillas, chimpanzees and bonobos, and us.

With respect to opsins and color vision, we humans are typical or "routine" trichromatic primates, having the typical arrangement of the single rod rhodopsin and the four cone opsins in our genomes, just like Old World monkeys and our other living hominoid relatives. So why call us trichromatic, if we have all four cone opsins? This is tricky. Most humans actually do have all four cone opsins—LWS, MWS, SWS1, and SWS2, and they should thus indeed be tetrachromatic. But what actually happens is that one of the two short wave opsins gets blocked by absorption and behaves as if there is only a single SWS opsin. More bizarrely yet, the "nonroutine" arrangement of opsin genes in New World monkeys results in a strange case where the males are dichromatic and the females are both dichromatic and trichromatic.

We hope this description of the opsins and their genes has not been too tricky to follow. But it is clearly evident that these genomic "wild cards" can produce color vision in many ways, and the story of the opsins demonstrates the plasticity of the evolutionary process when loci are as unstable as those for opsins are. In any event, once you grasp the opsin story, it is easy to understand how the arrangement of the cone cells in the retinas of vertebrate eyes leads to the plethora of colors that we and our primate cousins perceive. We will return in chapter 5 to how the rod rhodopsin and the four ciliary opsins are transformed by the brain into what we "see."

Sound Waves into Nerve Impulses

To understand how sound works, we need to know how our ears work. Once we have a good grasp of our own ears, we can then discuss how other organisms "hear." Our ears are made of an intricate set of structures that scientists divide into three areas—inner, middle, and outer. Let's start where the sound enters: the outer ear. This is composed of the big cartilaginous pinna (or auricle, the externally visible part of the ear), plus the external auditory meatus, or ear canal. Simply put, the pinna collects sound waves—vibrations—from the environment and funnels them into the ear canal, which in turn diverts the sounds to our middle ear, specifically the eardrum.

The middle ear is made up of the eardrum and three tiny bones called ossicles. The three ossicles are connected in a convoluted sequence. Connected to the eardrum is the first ossicle, known as the malleus (hammer), which in turn is connected to the incus (anvil). This is then connected to the stapes (stirrup), the inner end of which fits into a structure called the "oval window" of the inner ear. This completes the mechanical makeup of the middle ear. In contrast to the air-filled outer and middle ears, the organs of the inner ear are filled with fluid. They comprise three major structures: the cochlea (responsible for hearing) and the utricle and saccule (both responsible for balance). So this organ system supports two senses: balance and hearing. The cochlea, like the ossicles, is bony, but it has an interesting snail-like shape. Residing within its fluid-filled interior is the Organ of Corti, which is the sensory receptor for hearing within the cochlea. This organ is equipped with tiny hairs that transmit the impulses from external sound to the brain.

Sounds like quite a contraption. How does it work? Well, it all starts outside the ear, with a sound. Sounds are made of waves, which the ear has both to collect and to interpret. To do so, the pinna acts as collector, diverting the waves into the ear canal. The pinna also plays a role in determining how loud sounds are and from which direction they are coming. It is, as it were, the gatekeeper of sounds, but it is a very lax one because it lets all sounds in. Traveling along the meatus, the waves hit the eardrum. Aptly named (it's also known as the tympanic membrane), the eardrum acts to initiate the transformation of the sound waves into mechanical energy by moving back and forth in response to them.

Features of various sounds such as their volume, pitch, timbre, and rhythm will cause different vibrations in this membrane, leading to our ability to discriminate among sounds. Because the malleus is attached to the eardrum on the middle ear side, the mechanical energy from the vibrating membrane travels via this ossicle through the incus and on to the stapes,

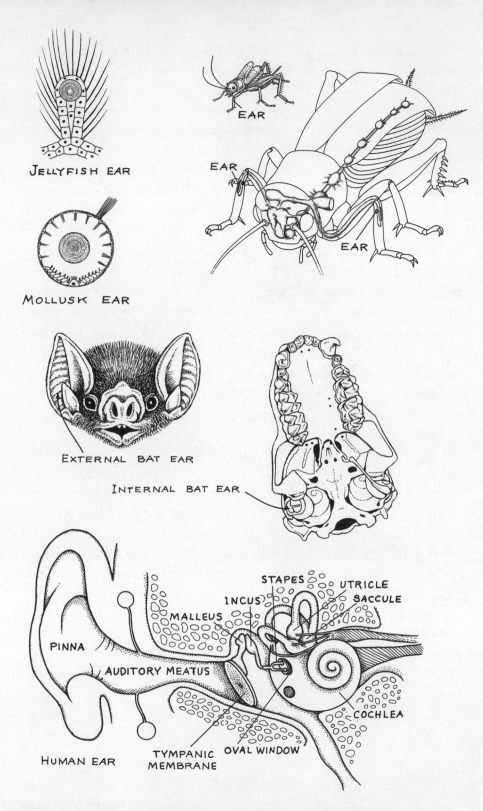

JELLYFISH EAR

EAR

EAR

EAR

MOLLUSK EAR

EXTERNAL BAT EAR

INTERNAL BAT EAR

STAPES UTRICLE
INCUS SACCULE
MALLEUS

PINNA

AUDITORY MEATUS

COCHLEA

HUMAN EAR

TYMPANIC OVAL WINDOW
MEMBRANE

The diversity of "ears." The owners of these "ears" are labeled in the figure.

which moves in an in-and-out fashion. These movements of the stirrup mimic the initial sounds that came into the outer ear, completing the transformation of sound waves into mechanical signals and transmitting them onward via its contact with the oval membrane at the entrance to the inner ear. This membrane is covered with tiny hairs that collect the mechanical movement and send it along to the brain. The movement of these hairs is very specific for different mechanical signals. For instance, if the mechanical energy coming into the cochlea carries with it information about pitch, it stimulates only a small patch of the hairs—and in a very specific way for each pitch the human ear can interpret. Other hairs will not be stimulated. Pitches we don't hear are the ones that don't stimulate any hairs in the cochlea.

Our brains are continually receiving information from the hairs of our cochlea; and one more transformation of the initial sound waves has to happen to get the information there. All of the hairs of the cochlea are connected to nerve cells, and the mechanical stimulation of those hair cells is turned into nerve impulses, which are sent to the brain via the acoustic nerve (the eighth nerve down in the cranium). Just as in seeing objects with the eyes, the ultimate message that the brain receives to interpret is a signal that has been transduced through cell-to-cell interactions from the ear—and, you guessed it, this happens via a G-coupled transduction reaction in the cells of the nervous system. The impulse then enters the brain and is processed (see chapter 5).

One Cannibal to Another, While Eating a Stand-Up Comic: "This Doesn't Taste as Funny as I Thought It Would"

The ability to taste is a hugely important sense, and animal behaviorists have long recognized that even very simple animals with small brains have food preferences. More than likely, though, the kind of sense of taste we have is a bilaterian invention. Insects and other invertebrates taste in pretty much the same way that they smell—through contact with molecules that are processed in their odorant system. The big difference is that smelling usually involves molecules that exist in the air, and taste involves molecules coming in direct contact with the animal itself; hence, taste is known as "contact chemoreception." That the sense of taste exists in simple animals such as leeches was established more than a century ago by an industrious naturalist named A. G. Bourne. In a quaint description of a simple experiment he performed in nature, Bourne was able to discern that leeches used some sort of sensory process (he preferred to call it taste) to detect the presence of a "meal":

> I picked up with my fingers a stone from the soft muddy bottom of a shallow, torpid stream. Returning to the same spot a few minutes afterwards, I noticed a number of leeches (apparently *Hirudo* sp.) swimming near the spot. On the following day, suspecting that they had "smelt" or "tasted" my hand in the water, I first stirred the surface of the mud with a stick, but no leeches appeared; after the water was clear again I "washed my hands" in the water without disturbing the mud, and very soon a number of leeches came up and swam about. The soft mud in which they live is about a foot deep, and although the disturbance of the surface mud with a stick was not sufficient to bring them out, the "smell" of my hands seems to have spread down and extended over an area of more than a yard.

In addition to tasting good themselves, some mollusks use their sense of taste to build their shells. Alistair Boettinger, Bard Ermentrout, and George Oster at the University of California at Berkeley have shown, using computer simulations, that the shapes and coloration of many kinds of mollusk shells can be explained by modeling what the mollusk tastes, using its tonguelike mantle. The mantle basically overlaps the edge of the growing shell. As the shell grows, the mollusk tastes the material of the shell—a simple calcium carbonate compound—from the day before. Because it can taste different concentrations of calcium carbonate and other aspects of the shell, it can use the tastes as a sort of memory of where to take up from the day before. This remarkable tasting process was incorporated into the model generated by Boettinger and his colleagues—and voilà! they could easily simulate many of the myriad mollusk shell shapes and color patterns observed in nature.

Closer to the natural world, scientists know that the simple thousand-celled nematode can taste salt, sweet, and bitter compounds. And it is well known that insects can also detect a large range of tastes. In general they do this through specialized organs in their mouths, but they have also developed some other very efficient and interesting ways to taste things that they eat or reproduce in. Some species of butterflies and the common house-fly can taste with their feet, because they have taste organs residing in those appendages. This arrangement of taste organs allows them to alight on a potential food, taste it, and decide whether they want to spend their time bending over and eating it. In some females of other insect species, taste organs exist in the egg-laying ovipositors. This allows them to taste the plant, fungal, or bacterial medium where they might deposit their eggs. If it tastes good, they lay eggs in it. Our major taste organ is of, course, our tongue. We wouldn't recommend you do this with anyone except someone you know well, but get up close and personal to someone's tongue. Take a close look, and you will see little pegs on its surface. If you can't see them clearly, put a little red wine or some blue Kool-Aid on the tongue, and these pegs, or papillae, will show up dramatically. Unfortunately, the little organs called taste buds that reside on them will not be visible. After the wine or blue Kool-Aid treatment, the papillae will stand out as little pale pink structures on a bluish background. If you can look really closely, the pegs should look a little like mushrooms, hence the name for them: fungiform papillae. Other parts of the tongue have different kinds of papillae; these are at the back and even on the epiglottis, so these areas can also transmit taste to your brain. An old wives' tale claimed that different parts of our tongues—that is, different clumps of fungiform papillae—were responsible for different kinds of taste. But it is now thought that the distribution of taste receptors is pretty equal across the tongue, so that all tastes are tasted equally with respect to location within the mouth.

Each taste bud is made up of around 50 to 150 receptor cells. These are clumped like a bunch of bananas and connected at their bases to a nerve fiber. At the other end of these elongated cells, facing upward from

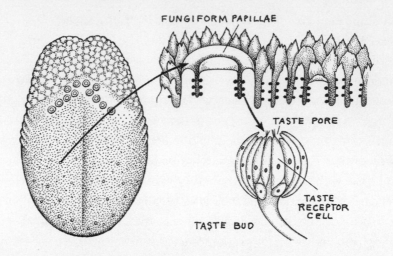

FUNGIFORM PAPILLAE

TASTE PORE

TASTE RECEPTOR CELL

TASTE BUD

A close-up view of the human tongue.

the tongue, is an area where the tips come together and form a little pocket called a taste pore. Little hairs (microvilli) from the taste receptor cell stick into the taste pore. The receptor proteins that detect the molecules eventually responsible for taste are found on these microvilli. Also connected to each taste receptor cell in these banana bunches is a nerve cell that is activated when the taste cell receptors react with specific kinds of molecules.

Many molecules will induce taste in this system. But there is only one kind of taste receptor per taste cell. There are five kinds of taste receptors—salty, sweet, sour, bitter, and a strange one called umami. The salty receptor detects molecules such as sodium chloride, and helps the brain modulate the salt balance in the body. If someone tastes too much salt—that is, so much salt that the brain decides "this will disrupt the salt balance"—then the individual will probably stop eating. The sweet receptor detects molecules like glucose and sucrose, and detecting this taste allows the brain to say, "Okay, nice, here comes some energy." Molecules with low pH (ones that are acidic) are detected by the sour receptors, and molecules like quinine trigger the bitter receptors in the taste buds. Those odd "umami" receptors detect amino acids like glutamate, yielding the typically strong tastes of broths, cheeses, and other concoctions with lots of amino acids.

Once the taste receptor reacts with a taste molecule such as an acid (for example, vinegar), how is this information transferred to the brain? Yes, once again G-coupled proteins are involved. But wait: G proteins are not the only way taste happens. For instance, while sweet, bitter, and umami do use G-protein-coupled receptors (GPCRs), sour and salty use different ways of getting the signal into the cell. Salty, caused by molecules like sodium chloride, simply works by reaction with ion channels that allow the sodium to move inside the cell. This movement of sodium ions into the cell depolarizes the entry area and creates an action potential that can then be transmitted to the nerve cells connected to the receptor cells. Sour acts through an ion channel, too, this one simply transporting into the interior of the cell hydrogen from the acids that cause the sour taste. This again disrupts the polarity of the cell and induces an action potential that is transmitted to the nerve cells connected to the sour taste cells. The hundreds of thousands of receptors on our tongues can be highly sensitive; some of our taste, though, is also enhanced by smell receptors and by the temperature of the tongue when the taste molecules hit it.

Odor

Like taste, smell is a sense that relies on detecting chemical information from the outside world. Both are thus known as chemosensory senses. Smells are nothing more than the sensations caused by our brains interpreting the presence, absence, and mixture of molecules floating in the air. That unpleasant smell from the garbage can outside occurs because microbes and other small organisms are degrading the garbage, creating emission compounds with sulfur and methane (among other chemicals). Smells can enter the bodies of organisms in many ways. For instance, a nematode smells with small sensory knobs that reside near its mouth. Nematodes are diverse, and the smelling organ in different nematode species varies greatly, but the general shape of the apparatus involves a hairlike "sensilla." The sensilla lies just below the outer cuticle and is made up of a socket cell that helps form the knob; a sheath cell that protects the knob;

and a neural cell that processes the odorant information. All three cells extend deep into the internal body of the nematode. Chemicals from the external environment are collected by the sensilla, and information about them is sent elsewhere in the worm's body. The level of detection that the simple sensilla can accomplish is so precise that, for example, when the common lab nematode *Caenorhabditis elegans* emerges from its larval case, it uses its sense of smell to determine how many other nematodes are nearby and how much food there is. Talk about body odor! If the food smell outweighs the body odor of other nematodes, the nematode stays and dines. If the body odor outweighs the smell of food, the nematode wisely moves on and forages elsewhere.

Fruit flies have yet other sets of organs for smelling. They primarily use their antennae as odorant receptor organs, but they also use their maxillary

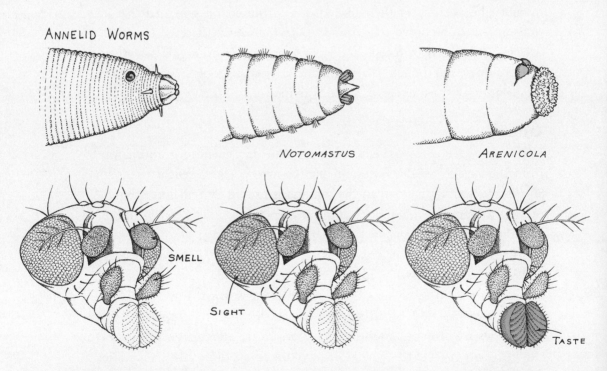

Nematode (above) and fly (below) heads showing the points of entry (shaded areas) for different kinds of external stimulae.

palps. These structures are knoblike protrusions that poke out and up near the proboscis (mouth). The maxillary palp and the antenna are made up of cells that include cuticles and a lot of little hairlike structures also called sensilla (the same word as in C. *elegans,* but maybe not the same things). The sensilla of the fruit fly contain olfactory nerve cells. In the simpler of the two organs, for each maxillary palp there are sixty sensilla poking out of the surface and two olfactory nerve cells per sensilla, for a total of 120 olfactory neuron cells on each palp. The twinned nerve cells respond differently to the chemicals they detect. Because there are two of them, there are three ways a smell signal can form for a fly. Cell one can be stimulated and cell two not; cell two can be stimulated and cell one not; and they can both be responsive. If they are both unresponsive, then there is simply no smell. This presents researchers with a computational way to look at smells in the *Drosophila* brain.

The third segment of the antenna carries most of the olfactory-processing cells for the fly. This segment of each of the antennae has about 1,200 olfactory nerve cells. The sensilla on the antennae are a bit more complex than those on the palp. There are four major kinds of sensilla: small basiconic, large basiconic, trichoid, and coeloconic. The most numerous are the basiconic sensilla, which have been studied in some detail. There are about sixteen kinds of olfactory neural cells that can be combined into pairs on each basiconic sensilla. Unlike the maxillary palp sensilla, this gives these antennal sensilla the ability to smell an incredible array of smells. With this many neurons, John Carlson and colleagues at Yale University point out that "in principle, if each ORN [olfactory receptor neuron] class represents a binary coding unit, the 16 classes described here could encode 216 binary codes, or some 65,000, different odors." The same researchers note, though, that the number of smells a fly can take in is most likely not nearly this large. The huge theoretical number is simply a result of the use of a combinatorial code of impulses.

The human nasal passage is a lot like the nematode's knobs and the fly's antennae. Our nasal receptor areas make up only about 2.5 square centimeters, but in this very small area there are over fifty million sensory receptor cells. Still, our olfactory system can only detect molecules with

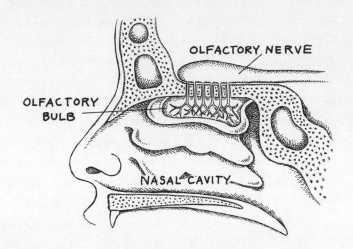

The human external and internal olfactory systems in cross section.

certain properties. The odorant molecules need first of all to be water-soluble. In essence, we are unable to smell anything in the air that is not water-soluble. Other properties that an odorant needs to have are high vapor pressure, low polarity (a chemical characteristic), and surface activity. Most odorants are also relatively small molecules. The odorants that our nasal passage collects then interact with receptors, tiny molecules that convey information to the brain. These receptors are on the surface of tiny hairs called cilia, and just how they actually detect and transmit information is a still a point of contention.

If we could look into our noses, we would see that the surface of the nasal area (also called the nasal epithelium) is lined with cilia that are jammed into the epithelium in a very thin layer of mucus (about 0.06 millimeters total thickness). The mucous layer is important, because it bathes the cilia in fluid, and makes the contact of odorants with the cilia more efficient. The mucous layer is made up of lipids that are important in transporting the smell molecules throughout the nasal passage, since the odorants we smell are soluble in this lipid mucous layer, giving them the ability to interact with the ciliary receptors. From each olfactory receptor neuron extend little hairlike cilia that look like whips, on which reside the receptor proteins that interact with the olfactory molecules. These olfactory neurons come together in what are called glomeruli, which are collections of ten to one hundred cells that converge together. Groups of glomeruli themselves converge into "mitral" cells, which reside in the olfactory bulb of the brain. The mitral cells are rather like relay centers that allow the olfactory bulb to communicate with the rest of the brain (see chapter 5). At this point we should note that there is a second kind of smelling mechanism in vertebrates that also conveys interesting and important smells and sensations to our brains. This alternative route is called the trigeminal sense, and it is considered olfactory because it involves neural cells embedded in the olfactory epithelium we have just described. These "tastelike" sensations include the tingly sensations one gets from some onions or the hot and cold sensations one gets from ingesting foods such as hot mustards and peppers or cool-tasting menthols.

We mentioned earlier that how the smell receptors are triggered has become a point of contention in the study of olfaction. When a chemoreception relationship exists between a smell molecule and its receptor,

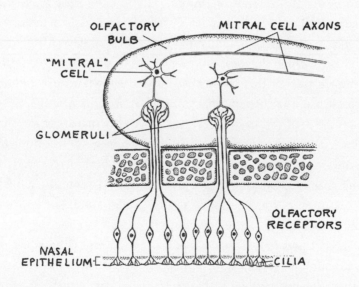

Each olfactory receptor neuron extends out little hairlike cilia that look like whips, on which reside the receptor proteins that interact with the olfactory molecules.

we assume it happens because the smell molecule physically interacts with the receptor. This somehow changes its conformation, and in turn this does something in the odorant receptor cell. What happens next? GPCR! More specifically, odorant or olfactory-type GPCRs connect to the cAMP pathway we've heard so much about before. The assumption here, though, is that there is physical contact of the smell molecule with the receptor and that the trigger is induced by physical contact. But Luca Turin, a biophysicist now at MIT, has suggested (following the work of Malcolm Dyson in the 1930s) that smell molecules vibrate when they are near the active sites of the receptor molecules. This vibration allows electrons from the smell molecule to "tunnel" through the molecule and to trigger the receptor to induce the GPCR on the inside of the cell. Because the smell molecule doesn't need to make contact with the receptor and because there is no real conformational change of the receptor, this mode of activating a receptor is known as the "swipe card mechanism." This contrasts with the usual way of thinking about how chemoreceptors work, which is more like a "lock and key mechanism."

According to Turin's way of thinking about odor reception, it works more like vision, in which light of different wavelengths interacts with various opsins. In Turin's world, smell molecules all have different vibrational qualities, which the odorant receptor molecules can detect. Although Turin's resurrection of this interesting idea about olfaction could actually make great sense, the conventional thinking is still that the GPCR connection makes the lock and key mechanism a better explanation for olfaction. And to date, all GPCR-mediated cell signaling appears to be implemented by the conformational change of the membrane-bound receptor protein, suggesting that card-swiping might not fit with G-protein-coupled reactions. What's more, experiments by Andreas Keller and Leslie B. Vosshall at New York's Rockefeller University showed no support for the vibrational theory. Vosshall and Keller used a specific prediction of the vibration theory, several human subjects, and some pods of vanilla to show that the vibration theory is not tenable or that, at the very best, it cannot explain all aspects of how odor detection works. But they do point out at the outset of their experiments that "at present no satisfactory theory exists to explain how a given molecule results in the perception of a particular smell." Vosshall explains that, although the current evidence is consistent with the lock and key hypothesis, the evidence doesn't *prove* that it is the mechanism by which odor detection occurs. So we will end our description of odor perception and simply stop and sniff the roses.

Touch

Unquestionably the most basic of all of our senses is that of touch. In its absence, we would truly be unable to feel part of the world around us. And to understand this vital sense we need to understand the largest, most expansive organ of the human body: our skin. To start at the very beginning, most organisms have an inside and an outside. Some, like bacteria and Archaea, have simple cell walls that separate the inside of the body from its outer surface. No ambiguities here. But becoming a multicelled organism changes the story dramatically. Different ways of evolving "inside" and "outside" have been experimented with over the one and a half billion years of eukaryotic life. These different ways of making an inside and an outside have influenced the ways in which different orgasnisms perceive the immediately surrounding world that actually comes into contact with their body. With increasing complexity among organisms, involving an expanding number of cell types and tissues, came the need to sequester these tissues into different parts of the body. And in many cases the ensemble of different tissues needed protection from the outside world.

Plants have evolved three major kinds of organ systems: dermal, vascular, and root. The dermal system is the outer protective coating of the plant, and it serves as a "skin equivalent." And though we've seen that it is controversial whether plants have anything resembling a nervous system (see chapter 2), this dermal skin does communicate, via molecular interactions, with other tissues in the vascular and root systems. So-called lower animals, such as cnidarians, have evolved a different and rather primitive way of keeping the outside world from their innards. These animals are also known as "diploblasts" because they have two cell layers—the endoderm and the ectoderm—that correspond to the inside and outside of the animal. Remarkably, these cell layers are each a single cell thick and are separated by a gelatinous layer called the mesoglea. The most complicated layer is the outer layer (ectoderm). It is also sometimes called the epidermis, and it contains several cell types. The innermost region, also called the gastrodermis, is where the organism digests food. The epidermis, the protective outer layer, also has nerve cells embedded in it that transmit sensory information to other parts of the cnidarian body. Remember, though, that Cnidaria don't have a brain. As a result, this nervous system is more like a net that transmits the sensory information throughout the body, galvanizing other cell types such as the epitheliomuscular cells that implement movement, the glandular cells that secrete mucus, and the nematocysts that send out nasty stinging propagules when the cnidarian is overstimulated by an external force.

So-called higher animals with bilateral symmetry are "triploblasts" with three layers of organization: endoderm, mesoderm, and ectoderm. Certain kinds of triploblasts, such as arthropods, have evolved a hard chitinous outer surface that serves as a support system for holding their bodies up and for protecting their internal organs from the outside world. Other triploblasts, such as the molluscan snails and clams, have evolved shells for protection and support of their bodies. But there are also triploblasts that have evolved no discernable exoskeleton or shell; these include worms and the cephalopods (octopus, squids). Vertebrates have evolved yet another way to support their bodies (with an internal skeleton) and hence need a completely different way of protecting their internal organs—hence the need for skin. Thus, when looking for these diverse organisms' strategies for sensing physical contact with the outside world, we see enormous differences. For us humans, the first line of defense is the skin. Our skin is made of cells composed of several layers. The cells of the outer layer are called the epidermis and implement the sensing of the immediate area outside of the body, allowing us to perceive contact, temperature, pain, and pressure. Our somatosensory system perceives and processes these sensations, and it receives them from four kinds of major receptors. Pressure, vibration, and whether something is smooth or rough are all perceived by mechanoreceptors variously called Merkel's disks, Meissner's corpuscles, Ruffini's corpuscles, and Pacinian corpuscles. Merkel's and Meissner's organs are the most sensitive of these and are found in the outer layers of the dermis and epidermis in skin that doesn't bear hair. For most of us, that means the palms, the lips, the bottoms of our feet, and the tips of our nose and tongue. These organs help us to determine not only whether we are touching something but also how long it touches us, since Merkel's disks are what is called slow-adapting, whereas the Meissner's corpuscles are fast-adapting receptors. The brain balances the signals from these receptors to determine length of contact. Deeper in the dermis, in joints, tendons, and muscles, are found the other two receptors—Pacinian and Ruffini's corpuscles—which for the most part sense vibrations. Without these mechanoreceptors, we would not be able to sense if our joints or muscles

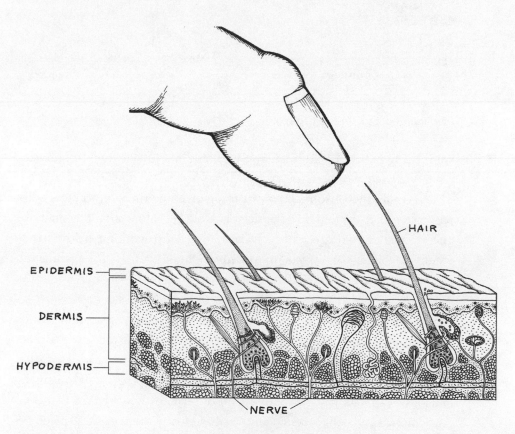

Diagrammatic representation of the touch organ—our skin.

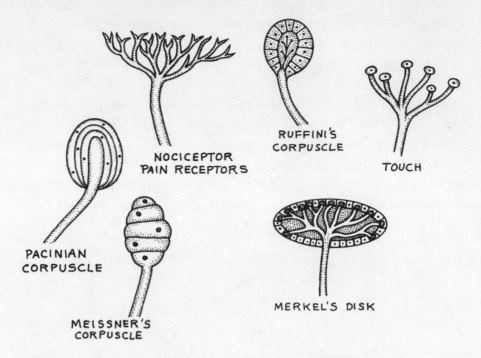

Mechanoreceptor cell structures.

were being pulled to extremes, and we could not judge when to pull back when we stretched them too far.

A second kind of somatosensory receptor is the thermoreceptors—hot and cold—that are found in the dermis. Cold receptors should actually be called "sort of cold" receptors, because they stop working below 41° F, when numbness usually sets in. At extreme temperatures, for that matter, hot receptors should be called "pretty hot" receptors, because they stop working at about 113° F. Below "sort of cold" numbness occurs, so all receptors in the sort of cold range fail to send impulses to the brain, which thus registers no sensation. But if the temperature drops too low, pain occurs because a receptor called Nav1.8 kicks in. This receptor detects extremely cold temperatures and resides in—of all places—pain-sensing neurons. Scientists who study this kind of receptor have shown that both cold-tolerant and cold-blooded animals have many more of these receptors in their dermis than warm-blooded, warm-tolerant animals do. Similarly, above the "pretty hot" range there are receptors in pain-sensing neurons that will signal the brain in the same way Nav1.8 does for very cold sensations.

Speaking of pain neurons, a third category of somatosensory receptor consists of what are formally called nocireceptors, or pain receptors that are triggered when skin, joints, or muscles are damaged. Our body has more than three million of these receptors distributed all over us externally and even in some internal organs. They transmit two types of pain to the brain. One is a sharp, instantaneous pain that signals your motor system to withdraw the hurting part of your body from the immediate stimulus to avoid further damage. The other is a duller background pain that tells your brain which area of the body has been injured and to temper its further use. The final kind of mechanoreceptor is the proprioceptors. These receptors sense the position of the body in space and the position of parts of the body in relation to other parts. They are what allow us to fine-tune our muscle movements by detecting how stretched and tense our muscles are so that we can modulate radically different tasks, such as sewing with a needle and thread versus driving a railway spike into the ground.

All of these receptors are molecules. One category of mechanoreceptors whose role in somatosensation is well understood is the transient receptor potential (TRP) channels. These channels are a diverse class of receptor molecules that lie in the membranes of cells. They are also used in smell and taste, and their role in sight and hearing is currently debated. But for somatosensation these molecules are supremely important. They have a deep phylogenetic history and are found in fungi and nonvertebrate animals such as *Drosophila* and *C. elegans*. Like opsins, part of their protein structure spans through cell membranes. In fact, there are six regions of the protein that slip and slide through the membrane to anchor it solidly. They can be stimulated by coming into contact with things or by changes in temperature. What most likely occurs when outside forces come into contact with a transient receptor potential is that it changes its shape, allowing the channel to exchange ions from the outside of the cell to its inside. Scientists who study TRP channel genes suggest that they are important in regulating calcium and calcium ion concentrations in cells.

As we have seen throughout this book so far, regulating ionic concentrations is the hallmark of sending electrochemical signals from cell to cell via synapses. By regulating calcium, potassium, sodium, and magnesium concentrations in nerve cells their electrical potential can be controlled, and this is the very basis of how nervous systems work. Still, although the observations made about TRP channels as receptors for external input into organisms are interesting, we need to point out that, as their name suggests, these proteins are transient and so must interact with other important nervous system proteins that form membrane channels (that is, potassium and calcium channels). Transient receptor potential channels are thus not the end-all of how nerve cells work, but they are a good way for organisms to process mechanical signals like touch into the complex nervous system. Yeast (one kind of fungus) have a single transient receptor potential channel gene product that resides in the membranes of vacuoles. Since yeast use these vacuoles to store things like toxic compounds, they need to communicate with the vacuole, and it is the transient receptor potential protein that does this. In yeast this protein is mechanosensitive, so that when it is jiggled by a toxin it triggers a defense response in the vacuole, inducing an osmotic change (osmosis being the process by which the concentration of water is adjusted on the inside and outside of cell membranes). The transient receptor potential channels are involved in osmosis, and this allows the yeast cell to keep track of anything toxic inside the vacuole.

In the fruit fly, TRP channels have diverse roles in the senses of vision, olfaction, hearing, and mechanoreception. And they are particularly interesting, if only for the crazy names that fruit fly geneticists have given to mutants for these proteins. Names such as nan, nanchung, water witch, painless, and pyrexia have all been used to denote the kinds of mutants studied. Flies with the water witch mutations cannot sense humidity in the air, whereas nan flies can't sense when it is too dry. Painless is interesting, because flies with this mutation cannot sense heat—put one of these flies on a hot stove, and it would feel nothing until it started to fry. Pyrexia is an even more extreme version of a heat-insensitive transient receptor potential protein. Flies with mutations in this gene become paralyzed when they are too hot (usually at temperatures in excess of 40° C).

When we move on to how senses, especially somatosensory perception, work in animals with backbones, we are faced with the same kinds of problems we considered in chapter 3. When we see an eye in a fruit fly, is it the "same" structure as the eye in a human? Even though both the eyes of flies and eyes of humans have rod and cone cells and use rhodopsin and other opsins to facilitate vision, in flies, transient receptor proteins are involved with the rhodopsin molecules in implementing vision. In humans (and other mammals), in contrast, transient receptor proteins are not interacting with rhodopsin. Curiously, though, transient receptor proteins do have a role in mammalian vision, because they interact not with the rods and cones (where rhodopsin and other opsins are happy) but with a completely different kind of cell called intrinsically photosensitive retinal ganglia cells. Through the action of transient receptor proteins, these cells tell the brain that too much light is coming into the eye, and the eye adjusts the pupil to compensate. So, while the transient receptor proteins in these photosensitive retinal ganglia cells don't facilitate the discernment of light wavelengths quite as transient receptor proteins in fly rods and cones do, they do have a role in vision. In a similar way, the other nonchemical, wave-based sense—hearing—also uses the transient receptor proteins in processing sound. And transient receptor proteins also have an important role in the chemical senses of taste and smell.

Let's now look quickly at the role of transient receptor proteins in somatosensory perception. As noted, these proteins are experts at detecting mechanical disruptions, and they somehow contrive to transmit a message back to the nervous system that a disturbance has occurred. In mammals, a transient receptor protein called TRPM8 is responsible for detecting menthol-like sensation on the skin. This protein is most likely the one that is active when someone gets a back rub with a menthol-based oil, or the one that was active back in our childhood, when Vicks VapoRub was splattered all over our chests for a cold. The VapoRub worked because the receptor was stimulated by the gradual temperature change caused by its menthol-based oils. Exactly how TRPM8 is activated and subsequently regulates the flow of calcium ions is not known, but we do know that it is not triggered by a chemical reaction. Menthol or menthol-like compounds somehow mechanically change the TRPM8, and this change allows the subsequent regulation of calcium ions across the membrane.

Mammal TRPM8 appears to have a close relative in flies that is called TRPM. Strangely enough, flies that have mutant TRPM genes have problems with a structure in their bodies called the Malpighian tubule, more or less the fly equivalent of the mammalian kidney. Nematodes also have proteins closely related to TRPM8, and these are active in the intestine of the worm. Does this mean that our ability to sense menthol comes from the digestive tracts of lower organisms? Actually not, because the mother of all TRPMs is actually that gene we discussed earlier called "painless"—the one that allows high temperature thresholds in flies.

The worm and fly genes that are more closely related to TRPM8 still modulate calcium levels in cells, though they do it in other tissues. Other transient receptor proteins that are of general interest include TRPV2, TRPV3, and TRPV4. Instead of being active in detecting cold like TRPM8, these channels are actually heat-detecting. It is interesting to note that there are no genes closely related to these transient receptor proteins in flies or worms. And what this means is that these genes expanded in vertebrates totally independently of what was happening in the invertebrates like flies and nematodes. Still, even though these channels work in a different manner from the systems we saw for sight, smell, taste, and hearing, it's worthwhile noticing that the initial outside influence (light, sound, chemicals, mechanical contact) is invariably converted into a nervous impulse of some sort.

Staying Oriented

Most organisms need to orient their bodies in space, and gravity is the force upon which the resulting awareness of balance is focused. Knowing in what direction gravity is working is the key to the vital knowledge of what is up and what is down. Even some bacteria need to orient themselves in space, although they don't use gravity to do it. The bacteria known as Magnetobacteria can directly orient on the magnetic poles of the earth by producing little magnets in their cells called magnetosomes. Why would a bacterium need to know where the poles of the earth are? Here's why.

Magnetobacteria don't like a lot of oxygen. So they prefer marine or aquatic environments where there are low oxygen concentrations, and they "swim" down into deep water to find them. If a cell were a magnetobacterium, it would be wasting time swimming left and right or back and forward. The only sensible direction to go is up or down, because the oxygen concentration will change more drastically in the up-and-down direction, and this will allow the tiny organism to explore the oxygen concentrations more efficiently. The magnetosomes allow the bacteria to orient North-South, so that they are parallel with the earth's surface. This orientation tells them that the up-and-down direction is perpendicular to their magnetically induced position, and they proceed accordingly. This is about as close as it gets to balance in bacteria.

Animals without bilateral symmetry also orient themselves with respect to the outside environment. Since sponges, cnidarians, ctenophorans, and placozoans are marine animals, one way to do this is by using light. If a bacterium is in the ocean, certain wavelengths of light will be coming from the surface, because that is where light from the sun hits first. The surface is up, so light will orient an organism to the surface in a process called phototaxis. But that's not all. These animals also use a process called geotaxis, or the use of gravity, to orient what is up and down; this is the process that bilaterans such as us also use. Phototaxis turns out to be a very efficient way of orienting oneself, and numerous methods of exploiting gravity have evolved. Cnidarians use specialized organs called statocysts (also seen in comb jellies and some bilaterians). The statocyst is a primitive balance organ, with a mineralized bead called a statolith inside a saclike structure containing a large number of tiny hairs. When the animal accelerates forward, the hairs are pushed against the statolith. As gravity pulls the hairs in the statocyst downward, this deflects the hairs away from the statolith itself and activates a neural signal to the cnidarian's neural net that in turn results in a response by the animal to the change in orientation. The neural net can then send messages to the rest of the body to compensate for the acceleration and maintain balance.

Vertebrates like us have the same need for orientation that bacteria and lower animals do. If we can't tell what is up and what is down, we are lost.

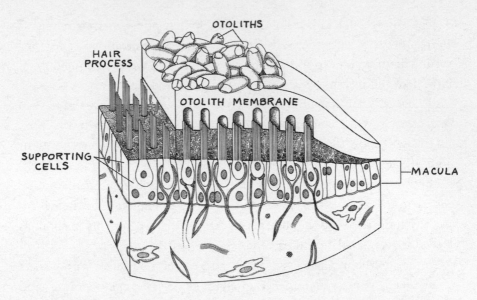

The cellular level of an organ of the vestibular system—the otolith organ.

Vertebrates have adopted a very specialized approach to keeping balanced. In the inner ear, in the same place where sound waves are collected for processing by the brain, lies the balance organ called the vestibular system. This system consists of three major structures—the semicircular canals (three of them), the utricle, and the saccule—and it lies close to the cochlea that we discussed in connection with the sense of hearing. Indeed, it is actually embedded in the same pool of fluid as the cochlea.

The utricle and the saccule work rather like statocysts and contain mineral specks known as otoliths (from the Greek *oto* = ear and *lithos* = stone; literally, "ear stone"). The utricle is like a carpenter's level placed through the ears. If the head leans left or right, the bubble in the level will move off center, and tiny sensory hairs in the utricle will come into contact with ear stones in the utricle. The saccule, in contrast, is like the same carpenter's level stuck through the forehead and out the back of the head parallel to the ground. Movement to the left and right will not move the level's bubble.

Although the utricle and saccule are responsible for detecting the position of the head in the vertical and horizontal planes, the three semicircular

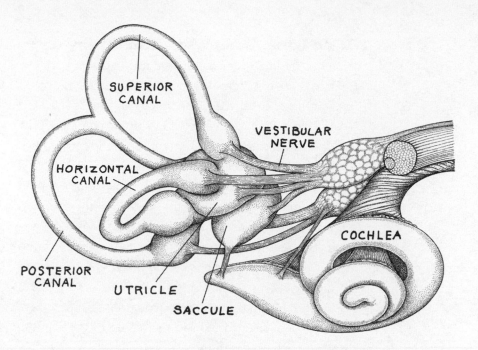

The inner ear balance structures, the three semicircular canals.

canals are responsible for monitoring the movement of the head. They are positioned in an X,Y,Z coordinate system with each canal lying parallel to each of the three planes in three dimensions. The three canals are known as the horizontal canal (X axis; tilting), the posterior canal (Y axis; up and down), and the anterior canal (Z axis; side to side). The fluid-filled canals are packed with tiny hairs in a bulb called an ampula, allowing the system to detect head movements. If the head is rotated in the horizontal plane, the hairs on the horizontal canal's ampula detect such movement because the hairs will slosh around in the ampula. Once the sloshing commences, the tiny hairs in the horizontal canal's ampula sends a signal to connecting nerve cells; and as some recent research indicates, G-proteins are involved in translating the mechanical stimulation into electrochemical information that can be used by the nervous system.

It's a Long, Long Way to the Brain

None of the many sensations we might detect with the different organs, hairs, or membrane receptors would make a bit of a difference if they didn't make it to the brain. What we have discussed in this chapter are all the mechanisms by which the external stimuli that impinge on our bodies and the bodies of other organisms are first treated. But to be of any use, all this received information has to be processed by the brain. And to get there, and to be intelligible to the nervous system, it needs to be exchanged into the currency of the nervous system, or into action potentials. In each case we have discussed, the stimulated receptor cell sends an "impulse" to the brain for further integration and interpretation. This process is true for all of the senses, and the nature of the impulse itself is also founded in molecular biology.

As we have tried to detail, G-protein-coupled receptors (GPCRs) are extremely important in this information exchange process. But moving the new currency around takes a dedicated road, which is provided by the nervous system. We have already introduced the basic units of the nervous system that shuttle action potentials from our sense organs to our brain and back out again. Many are involved purely in translation. For instance, say you are out walking. If you are leaning too far forward, the hairs in the saccule will detect the trend toward imbalance and trigger a response through a membrane receptor.

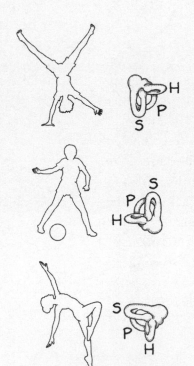

How balance works in our inner ear. The semicircular canals are drawn next to three poses of the human body oriented as they would be in these poses. The horizontal (H), superior (S), and posterior (P) canals take on different orientations in the different poses, and these different orientations are transmitted to the brain as indicated in the text.

The mechanical act of getting out of balance by leaning too far forward is then exchanged for an action potential via a G-protein-coupled reaction. This action potential is in turn transferred to the nerve cells connected to the saccule. And at that point the nerve cell connected to the saccule can send a signal to the brain, via the eighth cranial nerve, to the vestibular nuclei in the brain stem. Once this is done, the body can make a correction in muscle tone and speed to avoid falling.

But if you think things as easy as staying upright seem complex, imagine what we are in for when we look at even more involved perceptions of our outer world.

5 PROCESSING INFORMATION

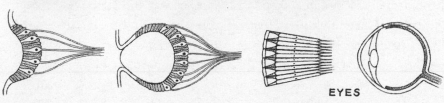

EYES

How do our brains take the action potentials they receive from our senses and exchange this currency of the nervous system to respond to environmental stimuli? In chapter 4, we looked at how our senses work at the molecular level. But as the action potentials make their way to the brain, we need to start explaining how this massive amount of information is processed into the things we see, touch, smell, feel, and taste. When we see something, someone, or anything with our eyes, what is our brain actually doing? Is the image being projected onto the back of our heads like some miniature drive-in theater? Certainly not, as we'll see. Still, how the sensory image is processed in our brains is only a small part of the story of how our brains evolved. Even very primitive animals, such as bacteria, protists (single-celled eukaryotes, like the deeming Dicty we discussed previously), and fungi, have ways of sensing, reacting, and orienting themselves to the larger world. But there's more to it than just sensing that outer world. Organisms need to interpret and react to it, and natural selection has ensured that their responses are relatively efficient. What brain processes have organisms evolved to handle their responses to the outer world?

The molecular reactions we discussed earlier (chapters 3 and 4) are the "first line of defense" in our reactions to the environment. Reflex reactions are perhaps our "second line of defense." Reflexes allow organisms to make rapid and efficient responses to challenges from the environment, such are as posed by predators and heat, as well as to imbalances in their physiology. So reflex reactions have evolved, above all, to be rapid and to involve as little of the brain as possible. In this latter respect they differ greatly from our emotional responses, another way in which we respond rapidly and efficiently to stimulation from the environment. Emotional responses involve memory, which is critical for our survival; almost daily there are reports of more discoveries about the importance of memory and how it is produced. What is more, our brains and those of other creatures have evolved to make decisions, and the processes by which we decide on our reactions to stimuli are yet another amazing part of the story.

How our brains evolved to do all these things is becoming clearer every day, not only as we learn more about how our brains actually function but as we are able to look in greater detail at our genomes and those of other organisms. The genes involved in brain processes can be compared among organisms, and because the genes carry their history along with them, the actual sequence of historical events can sometimes be retraced.

Single-Celled Organisms Can React in a Complex Manner to Sensing

In chapter 4, we assessed how the external forces around us get transformed into action potentials, which are then shuttled to our brains via our sensory nervous system. To follow up, let's ask, "How is this information processed in single-celled organisms and in simple animals?" As we've seen, even very simple organisms can detect light. This ability makes great sense, for most organisms are exposed to sunlight for a large part of each day, and being able to use the light as a cue, or as a landmark, has become an important aspect of how even very simple organisms survive.

Many bacteria have light-sensing cells that help them orient to light, and we have discussed this ability in terms of the Type 1 opsins such as bacteriorhodopsin and proteorhodopsin. It is clear that the genes involved in photoreception in bacteria are scattered among the millions of species of them out there. Archaea have light-sensing capabilities, too, but they do it a bit differently (remember that Archaea and Bacteria, even though both are single-celled organisms without nuclei, are fundamentally different kinds of organisms). One of the more astonishing light-sensing capabilities among bacteria is found in a strange group of microbes called Cyanobacteria. These bacteria belong to an ancient group that has evolved the ability to photosynthesize light into energy, much as plants do. As a matter of fact, very much as plants do, because it is generally thought that the organelle in plants that does the photosynthesizing (the chloroplast) was actually derived from cyanobacteria.

As eukaryotic cells were diverging to produce the many protist lineages we see in nature, along with the major groups of eukaryotes such as plants, animals, and fungi, some strange cannibalistic things were happening. The common ancestor of most organisms that possess chloroplasts (such as algae, plants, and a strange little protist called *Euglena*) apparently engulfed a cyanobacterium without digesting it. The cyanobacterium then engaged in what is called an endosymbiotic relationship with this common ancestor, so that all subsequent descendants from the original "hungry" ancestor actually possess the hitchhiking cyanobacterium-like endosymbiont in their cells. What's more, the endosymbiont retained its photosynthetic functions, including its genome! This may actually have happened more than once. For dinoflagellates (the first real "Dinos"), an interesting group of single-celled organisms with chloroplasts, are not closely related to plants, and the genes in their chloroplast are arranged bizarrely. It is quite possible that the ancestral Dino swallowed a different species of Cyanobacteria than did the ancestor of the plants and algae. And a group of algae known as the Cryptomonads apparently shares a common ancestry with a protist lineage without chloroplasts, so that the chloroplast in this lineage must have come from swallowing a related alga that did have chloroplasts. Apparently there are many ways to obtain chloroplasts, and one of the best is to dine on a species that already has them.

But back to our story about Cyanobacteria and sensing. When bathed in light, cyanobacteria will actually change colors differently when exposed to different kinds of light, in a process biologists call "acclimation." If the bacteria are growing in green light, they will change to a brick red color.

If the cyanobacteria detect that they are growing in red light, they will change to a bright blue-green color. These transformations do not just form part of a light show to attract mates or to warn others. Instead, the change in color implements a more efficient harvesting of light. The bright blue color of the bacterium is more efficient at absorbing red light, which occurs nearer the surface of aquatic or marine environments; the red color is more efficient at harvesting green light, deeper in the water. This response is unique because, while most single-celled organisms and even some simple animals can detect light, they cannot respond to it. It contrast, it appears that Cyanobacteria are similar to the human tanning fanatics we see today: they like their sunlight, and they ensure that they get as much of it as they can.

Cnidarian Cneural Nets and Sensing

In organisms with discrete nervous systems, we can start to examine how the brain is wired to allow the complex set of reactions to the outer world that is denied to simpler creatures. The simple animals Placozoa and Porifera have opsin genes, but how they use these genes, and if they really have a sense of sight—light detection—is difficult to say. None of the four or so placozoan cell types, nor the eight or so sponge cell types, has a light-sensing function. On the other hand, some Cnidaria do have light-sensing organs called ocelli ("little eyes") located at the base of their tentacles. But to say that a cnidarian sees is to stretch the metaphor. Basically, these light organs use opsins to gather the information about light in the outside world and send signals to the neural net. (Remember, cnidarians don't have brains, so the cnidarian neural net cannot form a real image.)

The same system, or something similar, operates in the many types of worms discussed in chapter 4 that have eyespots, eye patches, and light-sensing cells. We hope our earlier discussion of these organisms will have convinced you that, even though sometimes a worm is a worm, sometimes a worm is a worm is *not* a worm. But these organisms also have clumps of neural cells called ganglia, which form at the "head end" and in which information from the eyespots or light-sensing cells is processed. And the really interesting discovery about them is that the development of these organs, at least in some species, is controlled by many of the same genes that control eye development in higher animals like ourselves.

There are two actions these animals can take after their eyespots assimilate the light. They can tell it is light, and this sensing ability impacts their circadian (daily) rhythm and orients their bodies relative to the light (usually to run from it). This reaction is pretty simple and straightforward. But as eyes properly defined become more complicated, the sense of sight becomes more complicated. With actual eyes, processing the light signal involves very different neural pathways. In essence, animals with eyes can interpret incoming light as images—an ability that involves very different brain structures and functions. Only a few animal groups—the chordates, arthropods, and mollusks—have the biological capacity to resolve images in this way.

This actually seems like as good a place as any to ask the inevitable question about whether animals with nerve nets, like Cnidaria, or with quasi-brains, like platyhelminths, actually qualify as conscious? After all, they can detect and respond to stimuli. But the very question is a loaded one. We can, of course, attempt to set up rules for whether or not an animal is conscious. But as our children always remind us, rules were meant to be broken. In addition, our rules might not necessarily be yours. Yet we can ask some specific questions about consciousness and at least begin to understand the complexity of the problem. These simple animals are certainly aware of their outer world, but then so also are single-celled organisms. If this is your criterion, then single-celled organisms such as bacteria are by definition conscious. But simply sensing and responding to the outer world is not the same as responding in a directed and fine-tuned manner. Cnidaria tend to respond in a vaguely directed manner and in a kind of fine-tuned way. But one group of Cnidaria, the box jellies (those nasty little killers we considered in chapter 3), actually have pretty complex light-sensing organs and can use the information they obtain from these organs to track down prey in a very directed manner. This development represents a big advance over what the more primitive sponges and placozoans can do.

So what would the next advance in cognition be? One of the reasons that brains evolve is to make the organism more efficient and better able to handle fluctuations in the environment. The next step, then, might be to incorporate previous experiences into your responses, or perhaps even to be able to make decisions—rather as we have seen slime molds seem to. Or it might be to evolve mechanisms that make the "decision-making" process really rapid. In either event, it is not too difficult to see that the way to a "better" or more complex brain would be to increase the number of cells it contains. And it turns out that this trend toward increasing the number of brain cells is pretty much the strategy that we find as we trace the evolution of brain complexity throughout the rest of this chapter.

Insect Brain Complexity and the Senses

Let's go back to the arthropod brain, which often appears to be quite complex, and look at how it interacts with the sense organs. The prize for the most frequently studied arthropod brain has to go to *Drosophila melanogaster*. At ten thousand cells, the fruit fly brain has several orders of magnitude fewer brain cells than we do. But that doesn't mean a fly brain isn't complex. A feature of the arthropod brain known as the "mushroom body" provides an understanding of its developmental gene expression patterns. This lobed structure does look a bit like a mushroom, and it is connected to the brain's olfactory lobes. Its function, as Nicholas Strausfeld of the University of Arizona suggests, is to "conditionally relay to higher protocerebral centers information about sensory stimuli and the context in which they occur." Or perhaps "they play a central role in learning and memory."

Raju Tomer and his colleagues at the European Molecular Biology Laboratory used a battery of developmental genes to demonstrate that the polychaete worm *Platynereis dumerilii* also has mushroom bodies and that the very same genes are expressed in the same place in the mushroom bodies of both this polychaete and the remotely related *Drosophila*. They also suggested that the structure in the mammal brain known as the pallium shows similar genetic underpinnings to the mushroom body in invertebrates. The pallium of vertebrates is simply defined as "the layer of unmyelinated neurons [the gray matter] forming the cortex of the cerebrum," and remarkably, based on gene expression patterns, Tomer and colleagues claim homology of the cerebrum in the vertebrates with the mushroom bodies of invertebrates. An interesting hypothesis, but one we might nonetheless have to toss into our metaphor basket. Still, it is evident from this work that the mushroom body is an important area of the fly brain and that it receives and processes information from the sense organs.

With respect to sight, some arthropods (chelicerates and the strange organisms called velvet worms, or Onychophora) have what is called a central, or arcuate, body. This structure is most likely where vision is processed, at least initially. But it appears that for chelicerates (a group of arthropods that includes spiders, mites, and ticks) and onychophorans the visual input to the arcuate body is much more direct than the connections we find coming in from the eyes of insects, crustaceans, and myriapods (a group of arthropods that includes centipedes and millipedes). Still, it appears that that the neural structure of all arthropods is basically similar across the entire group, so it is a pretty good bet that the brains in arthropods are "the same" and that the arthropod method of processing visual signals arose early in arthropod history. This characteristic might actually reveal the role of cell number in generating the complexity of the brain because, in comparing brains of arthropods with those of vertebrates, it emerges that the greater the number of neural cells there are in an organism, the more complex brain function appears to be. All of the lower animals we discussed previously have extremely small numbers of neural cells in their body plans. But once we jump to organisms with hundreds of thousands and more cells in the brain, complexity and variability multiply, and fascinating things start happening.

Fish Brains, Fish Brains, Roly-Poly Fish Brains

We have already visited the brains of primitive chordates and their close relatives in chapter 3, but we purposely stopped just before getting to the description of the jawed vertebrates such as ourselves. Jawed vertebrates belong to a group called the gnathostomes, and the divergence of gnathostomes from other vertebrates is the first really big step toward the human brain. Primitive chordates had been around for quite a while before the jawed vertebrates diverged from the common ancestor of gnathostomes and all other chordates, and one of the major anatomical events that occurred in this transition was a dramatic increase in the amount of neural real estate among the gnathostomes. The common ancestor of gnathostomes appears to have experienced at least two rounds of whole-genome duplications, which means that the amount of genetic material in their genomes quadrupled! Many developmental biologists think that these genome duplications were critical in allowing the gnathostome genetic blueprints to expand and experiment. The basic set of genes could do the work of ensuring that the organism is fully functional, and the extra sets produced by the duplications could be "played with" to produce potentially important innovations.

Most of the early chordate lineages we reviewed in chapter 3 can be characterized as having the most rudimentary of brains. Some are nothing more than a patch of cells at the front of the animal. But the organisms we recognize as early fish begin to be brainier and have well-defined sections within the brain. Still, even so we can't say that at this time fish were the brainiest of organisms on the planet, because cephalopods were also evolving in this period. To explain another of those "phylogeny annoys me" situations, we should start by pointing out that, strictly speaking, either all jawed vertebrates are fish, or fish don't exist. Suffice it to say that the problem with the name "fish" as a biological group is the same problem we faced when we talked about protists. Fish seem to have done the same thing phylogenetically that protists did—diverge many times from a series of common ancestors. And some of the descendants (humans among them) have changed so much from the most remote common ancestor that we can no longer usefully describe them as "fish." Still, since all the things we think of today as fish share the same gnathostome ancestor and are relatively conservative in their body plans, their brains are quite similar from species to species.

Our American Museum of Natural History colleague John Maisey and several collaborators have reconstructed the oldest known fish brain, which they studied using the Grenoble Synchrotron, a powerful accelerator in France. This huge instrument allowed Maisey and colleagues to peer inside the cranium of a three-hundred-million-year-old fossil fish skull. Because the fossil was nicely preserved, the scientists were able to extract the outlines of its brain from its synchrotron image (this new technology is identical in principle to older techniques that used endocasts, or physical replicas of the interiors of braincases, to deduce the brain shapes of fossil organisms).

And it turns out that, like modern fish, this ancient specimen had a brain with three sections: midbrain, medulla, and cerebellum.

This tripartite (three-part) brain has a high degree of structural homology to other vertebrate brains. It has a "brain stem" that is connected to the rest of the nervous system of the fish body and that, as in other vertebrates, controls respiration and heartbeat. After the brain stem is the cerebellum, or "tiny brain," which helps coordinate precise, carefully timed movements that require constant adjustments based on visual feedback. The cerebellum is actually a very versatile part of the brain, because it has also been shown to be active in other, more abstract, brain functions. Along with the brain stem, the cerebellum makes up the hindbrain. Next comes the optic lobe, which in most fish is huge. This structure ensures that the fish can see really well, and as we saw in chapter 4, fish generally have great color vision. Of course, you might ask at this point, "What happens in fish that lose their eyes in the course of evolution, like the famous blind cave fish?" In 1872, F. W. Putnam became curious about this phenomenon and examined the brain of a species of blind cave fish. Lo and behold, the optic lobe in this species was as large as it is in fish with eyes. But about three decades later, a fish biologist named E. E. Ramsey spoiled the fun. He examined a second cave fish species and demonstrated the optic lobe and optic nerve were, as he put it, "measurably degenerate." Given there are very few species of blind fish on the planet, we can conclude basically "anything goes" in the optic lobe when a fish species loses its eyes. It is possible, for example, that Ramsey's species of blind fish lost its eyes a long time ago, whereas Putnam's lost its eyes more recently.

Continuing our journey through the fish brain, if you were to turn the fish on its back, you would see the part of the brain called the diencephalon. This part of the brain regulates hormone production. Fish even have a pineal gland, which is important in humans for sleep. But although in fish the pineal gland is used to process information about light, it also maintains the circadian rhythm of the animal. All of these structures (the optic lobe, diencephalon, and pineal gland) constitute the fish midbrain. Finally, at the front of the fish brain there are rather large olfactory lobes and a feature called the telencephalon. Both of these structures process signals from olfactory cues and are considered to be the last part of the tripartite brain.

At this point we should return to our initial task, which was to describe how the action potentials produced by the stimulation of the opsins in our rods and cones are translated by our brains into an image. This translation process actually varies a good bit among organisms with different kinds of eyes. In our case, though, several important electrochemical events come together to allow us to "see." Your eyes, with their rods and cones and opsins, respond to patterns of light by creating thousands of action potentials that originate in receptors that are at defined positions in your retina. These electrical signals are then sent to your brain via the optic nerves. In this way, one set of brain cells takes the signals in and relays them to other neurons in your brain, and on down the line. The neurons that receive these signals have all been specialized to interpret a specific aspect of the light bathing your retina, and the interaction of all of these incoming signals to your brain puts together the scene that you "see."

Let's use an excellent example from the American Museum's *Brain* exhibition to show how vision works. The example in question is an art piece created by Devorah Sperber, an artist who makes sculptural murals using spools of thread. Her artwork is fascinating because it toys with visual focus, imaging, and acuity. The artwork in the exhibition is a piece with 425 spools of thread of different colors. And it doesn't resonate much to the naked eye. But when you look through an optical device that flips the image upside down and flattens it out, Leonardo's Mona Lisa appears in sharp detail. How does a bunch of apparently oddly placed spools of

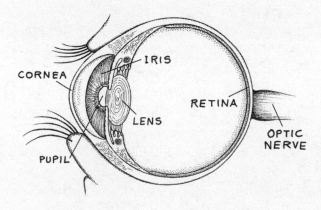

The eye, showing the lens, the cornea, the pupil, the iris, the optic nerve, and the retina.

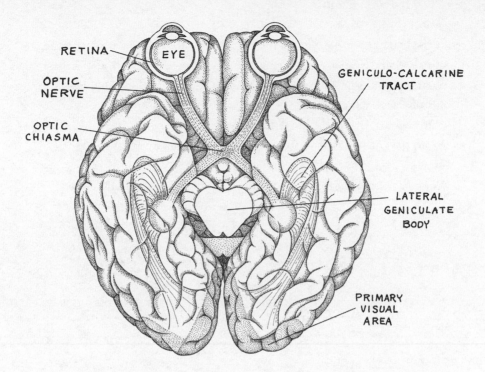

The human brain and optic nerve system.

thread convert to the Mona Lisa? In the words of the exhibit writers, seeing consists of several steps, and several parts of the brain are used to implement it:

> Seeing may seem like a simple thing, but it doesn't happen in just one step. Many parts of the brain must work together.

PICKING UP LIGHT
Light waves hit the *retina* at the back of your eyes and are absorbed by cells (rods and cones) that convert them into action potentials.

RELAYING SIGNALS
The optic nerve collects the signals and carries them to the *thalamus,* a dispatcher for all kinds of sensory information.

PROCESSING SIGNALS

In the *visual cortex* at the back of the brain, specialized neurons receive the signals. Some neurons detect specific visual elements like lines, shapes, color, and motion. For the spool-thread Mona Lisa these include:

Detecting lines at different angles: the elements of Mona Lisa from horizontal lines, vertical lines, and forty-five-degree angles

Sensing color: of the Mona Lisa where a certain color appears

Spotting a figure: silhouette of Mona Lisa

Recognizing a face: contour and specific landmarks of Mona Lisa's face

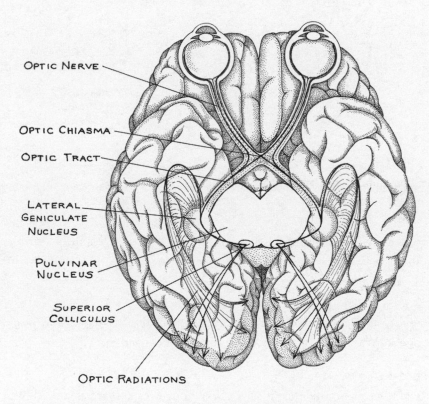

OPTIC NERVE

OPTIC CHIASMA

OPTIC TRACT

LATERAL GENICULATE NUCLEUS

PULVINAR NUCLEUS

SUPERIOR COLLICULUS

OPTIC RADIATIONS

Visual circuitry of the brain. The external light stimulus enters the eye, interacts with the retina, and is converted to an action potential that travels down the optic nerve, where the action potential crosses from one side of the brain to the other via the optic chiasma. The information then travels to the lateral geniculate nucleus and on to other regions of the brain such as the superior colliculus and the pulvinar nucleus, as indicated by the arrows (optic radiations) to the primary visual area.

FINDING MEANING

Other brain areas linked to memories, thoughts, and emotions help you understand what you see.

This description effectively conveys the overall process by which vision proceeds, using distinct circuitry in the brain that psychologists and neurobiologists have deciphered over the past couple of decades.

And although the pathways that have been defined in these studies are complex, one common factor is that they are traversed by action potentials, which stimulate different neurons in the pathways and optic areas to result eventually in the images we "see." The most spectacular thing about this system is that the vision which gives us such a clear view of the external world in which we live is neither a singular process nor the result of a singular impulse in the brain. What we experience as vision is instead the interactive sum of many impulses in many different parts of the brain.

No Pain, No Gain

Note that in fish there is no structure we can call a cerebral cortex or "neocortex," a major feature of our own brains. This lack of a neocortex has sparked an interesting debate about whether fish feel pain. Some researchers are adamant that, since fish don't have a neocortex, they can't feel pain, because the neocortex is where the neural impulses from heat, cold, or mechanosensory stimulation are processed in the brain of higher vertebrates. When considering pain, we first need to recognize that the notion of pain should be kept separate from that of "nociception" (chapter 4), which is reflexive. If you don't have the part of the brain that processes pain, the argument goes, then you can't feel pain. You do feel the outside environment, but it is not painful as we understand it.

In a spectacularly mean experiment, some researchers at Stanford University decided to test whether fish experience pain. They mixed up some acid and honey bee venom and injected these noxious compounds into the lips of several test fish. The behavior they observed was, they thought, an indication that the fish were indeed experiencing pain: the fish rubbed their mouths on the floor of the aquarium, as if trying to sooth their aching lips. Still, James Rose of the University of Wyoming rejected this interpretation, arguing that without any neocortex fish don't have the neural make-up to experience pain. And he also pointed out that the researchers had pumped ridiculously large amounts of acid and venom into the lips of the unfortunate fish, arguing that they would have reacted much more violently if they had pain receptors and had actually experienced pain. In Rose's view, the work claiming pain in the subject fish actually did no more than demonstrate that they had nociception. It is no surprise that Rose is not adored by animal rights activists, who claim he is biased because he likes to fish. After all, he lives in Wyoming, where some of the best fly-fishing in the world can be experienced.

Can looking at other groups help resolve this controversy? For example, do flies feel pain? If Rose's principal criterion is used, we have to say "no" because, even though some scientists think that there is a tripartite organization in some invertebrate brains, these creatures do not seem to possess an equivalent to the neocortex in which pain is processed. Still, flies do have nociception: they will react to high or low temperatures as detailed in chapter 4, but these reactions can be considered reflex. Jane Smith, an ethicist at England's University of Birmingham, addressed the problem of invertebrate pain in some detail but ended by suggesting that "because pain is a subjective experience, it is highly unlikely that any clear-cut, definitive criteria will ever be found to decide this question." In other words, the subjectivity involved in defining pain will never allow us to say, "Yes, invertebrates feel pain." But it also means that we will probably never be able to say, "No, invertebrates definitely do not feel pain." Nonetheless, Smith also suggests that we can use certain behavioral responses to noxious stimuli as indicators that pain might exist in organisms like invertebrates.

All of this reasoning raises the question of why pain should exist at all in organisms with a neocortex. If the ubiquitous nociception is enough, why is another mechanism like pain necessary? One answer that has been offered is that pain might actually be adaptive. It would be a reinforcing mechanism for nociception, just to ensure that the organism responds appropriately to a noxious stimulus. Still, as we pointed out in chapter 1, claiming adaptation as an explanation for everything is problematic; and in this case, it doesn't get us very far. Perhaps, then, pain might be a by-product of the evolution of other traits.

Let's provisionally propose that fish and invertebrates don't experience pain. Following up on this one might ask, "Do they have emotions?" Well, it's known that fish can experience stress, but this introduces the same problem that we had with nociception and pain. Stress is at its root a physical problem, not an emotional one. Some think that fish can express happiness or sadness; this is a notion that will be easier to examine once we have discovered just where in the brains of animals the emotions are controlled. Let's first look at where the emotions reside in the brains of animals other than ourselves. After the emergence of fishlike forms, the next big evolutionary event was the divergence of a fish lineage that gave rise to the four-legged vertebrates called tetrapods.

Many of us are familiar with the Gary Larson cartoon in which two fish playing baseball in the water are gazing at their only ball sitting on an island. Truly, four supporting legs represented an amazing innovation for the tetrapod lineage, and it would eventually have allowed the fish in the cartoon to gather up their baseball if they had waited long enough. But four legs are trivial compared to the innovations that had evolved in the brain of their common ancestor. Specifically, two very important events occurred. The first was the evolution of what for the lack of a better word is known as the "limbic system," and the second was the layering of a sheet of cells—the "cortex"—over the basic fish brain.

"Limbic" and "Triune" Are Dirty Words, Too

To understand more clearly the next step in the evolution of the vertebrate brain, we need to return to the fish telencephalon that, as we mentioned, forms part of the forebrain. The forebrain is made up of two important components, and as the area from which our vaunted cerebral cortex originated, it is by far the most interesting (to us) of the three major brain regions in the fish brain. Early studies of the fish telencephalon and its homologous regions in other vertebrates suggested that the major parts of the forebrain had appeared in sequence. Specifically, a part of the telencephalon called the globus pallidus was thought to have arisen in the common ancestor of all gnathostomes. Another part of the forebrain known as the striatum (made up of the caudate and the putamen) was thought to have evolved next, in the ancestor of amphibians and the rest of the vertebrates. And the cerebral cortex itself developed later, with the common ancestor of birds, reptiles, and mammals.

The striatum and globus pallidus constitute what most neurobiologists now call the basal ganglia. The basal ganglia most certainly control movement, if not a bunch of other functions as well. If the basal ganglia are deficient or injured, motor movement is impaired (Parkinson's and Huntington's diseases are two examples of such impairment). In these older ideas about the evolution of the forebrain, it was thought that the cerebral cortex had taken over some of the functions of the basal ganglia as mammals evolved. The striatum and globus pallidus were thought in turn to be responsible for the primitive kinds of movements that are stereotypical of reptiles. Because the cortex was believed to have taken over control of finer motor movement in mammals, researchers believed that they could explain why mammalian movement was so much more adaptable to the environment than, say, a reptile's. It turns out, though, that this serial way of thinking about the basal ganglia is flawed.

One way to examine the situation is to ask, "Can you have a vertebrate brain without the basal ganglia?" Why? Well, if the cortex supplanted functions of the basal ganglia, the latter might not be necessary for the complex activity of the vertebrate brain. But in a paper with one of the most definitive titles we have ever seen, Anton Reiner, at the University of Tennessee, answers resoundingly: "You CANNOT have a vertebrate brain without a basal ganglia." His reasoning for answering so emphatically is that, after detailed examination of several kinds of vertebrate brains, "the striatum and pallidum have been the basal ganglia constituents since the earliest jawed fish." He also argues, on the basis of the "wiring" of the various neurons involved in the basal ganglia, that removing these ganglia would be like removing a relay station in an electrically wired system. In his view, the basal ganglia were required to ensure the proper wiring of the vertebrate brain, even as the limbic system and the cortex evolved in the common ancestor of mammals.

Some neurobiologists do not like the term "limbic system" because it is vague. It loses its definition under scrutiny because it consists of an agglomeration of several more precisely defined brain regions, and there is little agreement about its real limits. Where most neurobiologists concur is that three of the many localized brain regions in the general area of the limbic system are the amygdala, the hippocampus, and the hypothalamus. Other brain regions may also be included in the limbic system, depending on the definition you are using. These regions include the thalamus, the

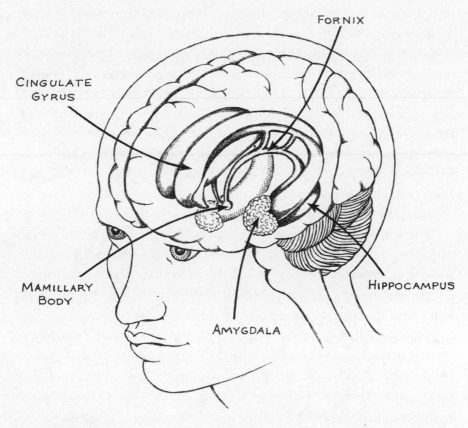

FORNIX

CINGULATE
GYRUS

MAMILLARY
BODY

AMYGDALA

HIPPOCAMPUS

The "limbic system."

pituitary gland, and even the olfactory bulbs. So let's ask what, if anything, a limbic system might reasonably be.

One way to look at this question is to look at the history of how the term has been used. From this we can draw two conclusions: (1) referring to a limbic system implies that mammals have a discrete part of the brain dedicated to the emotions; and (2) the limbic system evolved only in mammals and therefore is a character that existed in the common ancestor to all mammals. To address the first point, Mark Bear, Barry W. Connors, and Michael A. Paradiso, authors of a highly successful neurobiology textbook, point out that "given the diversity of emotions we experience, there is no compelling reason to think that only one system—rather than several—is involved. Conversely, solid evidence indicates that some structures involved in emotions are also involved in other functions." It is thus becoming clear to most neurobiologists that the "limbic system" tag oversimplifies both what our emotions are all about and how brains are structured. As for the second point, anatomical studies now suggest that some of the structures traditionally considered part of the limbic system are found among lizards and birds, as well as in mammals. The best explanation for these observations is that the common ancestor of lizards, birds, and mammals had in fact possessed these structures—a direct contradiction of the notion that the limbic system is a mammalian invention. And this issue brings us to bird brains.

When we think of all of the complex behaviors that birds demonstrate, we might well suspect that something interesting has happened in this major lineage with respect to the brain. First, if we calculate the brain-to-body ratio of birds, it turns out to be very similar to the same ratio in apes. Relatively speaking, then, birds' brains are not really that small to begin with. Equally important in understanding birds is their genealogy. This large and varied group of organisms (birds have managed to adapt to almost every ecological zone on the planet) shares a unique common ancestry with lizards and, more recently, with dinosaurs. Indeed, birds are often considered to be the only living group of dinosaurs on earth today. Specifically, they share a common ancestor with a group of dinosaurs called coelurosaurs, and they should thus be considered a highly derived dinosaur group. Because their brains are not too terribly similar to ours with respect to the cortex, birds have been rudely characterized as "bird brains" or "featherbrains." But in fact birds have more complex brains than lizards have. Everything a lizard has, a bird also has, and then some. They have a more developed cortex than lizards, making them in many ways more similar to mammals.

But because birds are considered highly derived reptiles, the similarities of bird brains to mammal brains must have occurred by convergence. Indeed, Nathan Emery and Nicola Clayton of Cambridge University have proposed that "there are clues from modern neuroanatomy suggesting that avian and mammalian brains may have come up with similar solutions to the same problems, thus demonstrating both mental and neural convergent evolution." The convergence is even more remarkable when one looks at the similarities inside the brains of birds and mammals. It turns out that birds and mammals have similar neural wiring relating both to vision and to vocal learning. What's more, birds do have what many neuroanatomists would call certain components of a limbic system. Still, because of their closer evolutionary connection to lizards than to us, the innovations that evolved in our lineage were obviously not a part of theirs. And equally, the original structures that were part of the bird-lizard ancestor's inner brain have acquired innovations that ours have not.

We tend to agree that when a definition of a structure becomes vague, as in the case of the limbic system, it loses power. The notion of a unitary limbic system will cause all kinds of problems in making sense of homology and will therefore confuse all metaphors that might actually be useful. Still, whether or not a neurobiologist abandons a term is dependent on how stubborn he or she is. Expunging a term from the lexicon of a discipline is very hard, sometimes impossible. A famous scientist suggested once that a term he didn't like was "ripe for burning." Guess what that term was? Homology! But even though this famous biologist wanted the term to be purged from the biological lexicon, it persists—because its use is essential in understanding that most basic of evolutionary principles, descent with modification. So while the term "limbic system" evidently has its uses and will most likely hang on, it is important that neurobiologists recognize its

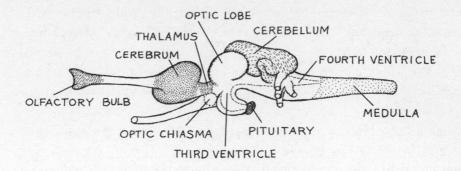

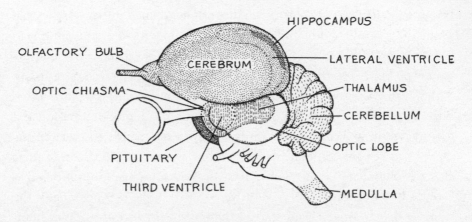

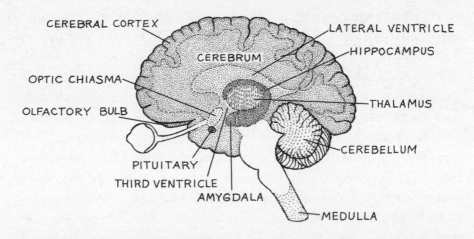

Comparison of lizard (top), bird (middle), and human (bottom) brains, showing the similarity in increase in complexity of the cortex in birds and mammals. Note that the lizard lacks the expanded cortex regions that the bird and mammal have, but note also the areas of similarity across all three brains as discussed in the text.

limitations and that it is most valuable in the context of evolutionary metaphor.

Another once-popular idea about how the brain evolved is called the "triune brain" theory. But it, too, has been under fire lately. The term was introduced to describe the three-layered brain that we humans have, and it should not be confused with the three-part—hindbrain, midbrain, forebrain—brain we have just been discussing. The triune brain was coined several decades ago by the Yale neurobiologist Paul MacLean, who meant it to describe the sequential addition of brain layers through evolutionary time. It's an idea rather like that of the sequential addition of layers in the basal ganglia. The triune brain started out with a lizard brain. Next came a "proto-mammal" brain, with a limbic system. And finally came the cortex, which evolved to prominence in mammals and is most greatly enlarged in primates. Although this explanation for the complexity of our brains is appealing and helps us to understand the structure of our brains from the inside out, it actually represents a gross oversimplification of how our brains evolved.

One reason this view of the brain was so appealing is that it follows an age-old way of describing nature that dates back at least to Aristotle. This is the *scala naturae,* or the Great Chain of Being, and it suggests that life can be thought of as progressing from the very simple to the highly complex. Starting with dirt, the Great Chain sweeps upward through plants, then through the lower animals, and on toward perfection via an almost infinite succession of forms. At the apex of the scala naturae are human beings, themselves transcended in the medieval Christian version of the chain by the angels and, ultimately, God. We hope it is easy to see why this way of thinking is flawed. As we pointed out in chapter 1, evolution does not proceed from primitiveness to perfection. More important, all organisms on this planet have equally long evolutionary histories. They must. All organisms on earth emanate from a single common ancestor, and so the bacteria on your teeth, the dog in your house, and the plants in your garden have all gone through massive amounts of descent with modification. No single organism is "better evolved" than any other.

You might, then, take issue with how we have sometimes taken a human-centered approach to understanding brain evolution in this book.

But our readers, we can be pretty certain, won't be bacteria or dogs or squids. If SpongeBob's cephalopod friend Squidward were the target audience of this book, then we certainly wouldn't have taken so much time focusing on the validity of the word "limbic" or whether there is a triune brain. Instead, we would probably have focused on the nuances of the cephalopod brain. To get to a cephalopod brain, some of the ancestors we have talked about so far would indeed have been relevant, so some of the story would have been the same. But after protostomes had diverged from deuterostomes, the story would have diverged. Still, whichever end-product we focus on, the principles of brain function and evolution remain essentially the same. Which means among other things that, like most of the imprecise terms that persist, "triune brain" has its utility in helping us understand how the parts of the brain fit together. Where it fails, is in describing how the brain evolved.

Only by examining actual ancestors, or ancestors reconstructed from comparative evidence, can we make sense of the evolutionary events that led to the emergence of our thinking selves. Lining up living forms and transforming them from one into the next is an appealing but flawed way to view our evolution. We simply cannot place ourselves at the top of a sequential series and learn anything that is useful about how we got that way. Still, we cannot doubt that we humans have many interesting behaviors that were acquired from a series of now-extinct ancestors. But before we move on to those subjects, we need to look at some of the molecular and physiological aspects of how these brain regions work.

Eat the Pudding, Eat the Pudding, Eat the Pudding

When you haven't eaten food in a while, what happens? You feel hungry. But what does feeling hungry really mean? It is simply the sensation of your body telling you that you need food. Think about feeling hungry at around four in the afternoon. Maybe your stomach is growling; perhaps you're a little light-headed and tired. This is your body telling you and Homer Simpson to "eat the pudding, eat the pudding, eat the pudding." It's a very physiological reaction to your situation. But how about when you are about to face something awful, like having to pay overdue bills or telling a friend something he or she might not want to hear? You feel anxiety. And this time the feeling emanates from somewhere else. It is beyond physiological: this time it is more of an emotional feeling telling you that you need to respond to something. The feeling might also affect you physiologically, like a queasy feeling in the stomach or sweaty palms; but the feeling itself is very different from hunger.

Anxiety and hunger are thus two divergent responses to stimuli you experience. One is visceral, the other more ethereal, and they are different in origin. Your physiology has created the feeling of hunger, but your brain has produced that feeling of anxiety. How does all of this work? Well, although this matter of feelings is complex, one good way to understand it is by realizing that to have the effects of producing anxiety or fear or happiness, the action potentials coursing through our axons need to be regulated (turned on and off, and up and down) and sent to the right places in our brains and peripheral nervous system. It is the combination of electrochemical impulses in a specific part of the brain that produces the emotional response. One of the cleverer interactives in the American Museum exhibition that this book is based on concerns our attempt to explain how neurotransmitters affect our emotions in everyday life. The interactive, developed by combining the scientific input of Dr. Maggie Zellner, with the creative instincts of Helene Alonso, director of interactive exhibits at the museum, follows what happens in the brain of a young person who is confronted with an everyday set of problems.

The interactive has two dashboards. The first is a set of four registers measuring levels of four kinds of neurotransmitters: dopamine, endorphins, oxytocin, and stress hormones like cortisol. The registers fill with different colors representing the four types of neurotransmitter. A brain map on this dashboard indicates the amount and position of each neurotransmitter. The second dashboard shows several physiological systems in the child's body—stomach, lungs, heart, muscles, and sweat glands—that will be affected by the activity of the surging and subsiding neurotransmitters.

As you approach the interactive, you are asked if you wish to play or to learn more about neurotransmitters. If you play the game, the interactive starts out with the kid sitting reading a book. The kid smells a cookie being baked. In the brain diagram on the transmitter dashboard, the olfactory region of the brain lights up, and then lots of dopamine is produced and floods the striatum, the hippocampus, and the prefrontal cortex. The dopamine levels rise on the dashboard meter. The neurotransmitter dopamine is a tiny molecule but is nonetheless a powerful player in the physiology of our brains; it is generally considered to be responsible for monitoring needs and cravings for things. With dopamine flooding the striatum, the stomach growls and the mouth waters on the physiology dashboard, in anticipation of eating those tasty cookies that Mom is baking. But a warning pops up in the interactive, because Mom has warned the kid, "No snacks before dinner." You can see a little blip of stress hormone go up on the transmitter dashboard. At this point, the interactive asks the player to choose: (A) Obey Mom and go out to play, or (B) Go get a cookie. We have yet to see a single person interacting with this exhibit make choice A. Everyone wants the cookie. Choosing B means that the kid's dopamine levels go up even more. In addition, in the brain diagram on the transmitter dashboard we see some activity in the kid's motor cortex, the strip of the cortex controlling movement. Once the kid is in the kitchen, the cookies are fair game, and when the cookies are sighted, the kid's dopamine level rises even more. But an image of Mom appears in a bubble on the screen. The hippocampal regions of the brain light up on the transmitter dashboard, and the kid remembers that Mom said, "No snacks before dinner!" Of course, the kid's stress hormone levels shoot up on the register on the neurotransmitter dashboard, while on the physiology dashboard the heartbeat quickens, the breathing gets more rapid, and the kid begins to sweat a little. When the kid grabs and eats a cookie, the pituitary gland immediately delivers endorphins to the bloodstream and the hypothalamus releases endorphins to several parts of the brain.

Now the kid is quite happy, and breathing slows along with the heartbeat. But then Mom appears in the doorway. "Busted!" says the interactive. Immediately the kid's stress hormone levels skyrocket on the transmitter dashboard. The physiology dashboard is going crazy: heartbeat high, rapid breathing, muscles tense, and palms sweaty. At this point the interactive asks: Do you (A) Stay and apologize to Mom, or (B) Run outside and avoid Mom? Most people figure they have been bad enough at this point, so they press A and apologize. Once the apology is made, the stress hormone level plummets. As the kid makes up with Mom, the oxytocin levels go way up and the endorphin levels rise. The kid's bond with Mom is reinforced with oxytocin, and the endorphins make him feel pretty darn good. The physiology dashboard now resets, and the kid is ready for the next set of challenges life has to offer. Visitors using the interactive leave with a pretty clear picture of how neurotransmitters lie at the heart of how these actions and reactions are governed in the brain.

As we've seen, neurotransmitters are small molecules that are released into the synapse area from vacuoles in the presynaptic neural cell. These molecules are diverse. They range from small compounds like nitrous oxide, carbon monoxide, and acetylcholine, to small amino acids like glutamate, to molecules called monoamines and including epinephrine (also known as adrenaline), to larger molecules like hormones, and to even larger chains of amino acids called peptides (small proteins). Neurotransmitters collect in those vacuoles of the presynaptic cell and are released as a result of action potentials or graded electrochemical potentials that reach the presynaptic inner membrane edge of the synapse. If the action or graded potential is of a proper degree and frequency, the vacuole merges with the inside of the membrane of the presynaptic cell, and—voilà!—a package of neurotransmitters is emitted into the synapse. The use of amino acids or of short chains of them as neurotransmitters more than likely came about because there are a lot of them around (as a result of the breakdown of nutrients), and they simply get recycled as neurotransmitting components of the synapse. But some neurotransmitters are specific to localized parts of the brain. This indicates that the cells in those regions are specialized for making or processing a specific neurotransmitter. Neurotransmitters act by binding specifically to a receptor molecule that is embedded in the cell membrane of the postsynaptic neuron facing the synapse. Binding of the neurotransmitter to the receptor produces a change in the receptor, and this triggers the kinds of intracellular reactions we talked about in chapter 3. Depending on the neurotransmitter involved and the region of the brain in which it is being produced, the postsynaptic cell will respond by being either inhibited or stimulated.

Specific neurotransmitters have specific jobs. For instance, glutamate, that small amino acid transmitter we've already mentioned, affects what are called fast excitatory synapses both in the brain and in the spinal cord. Glutamate is therefore an excitatory neurotransmitter, one of the most important ones in our brains. It is no wonder that plants target this neurotransmitter in developing defenses against being eaten. Glutamate is also active—indeed, the major player—at modifiable synapses, which can increase or decrease the strength of connection and impulse. Another ubiquitous neurotransmitter is the familiar GABA. This acts a lot like glutamate, except instead of being excitatory, it is inhibitory. It is the "off switch" for neural transmission, much as glutamate is the "on switch." As an example of a highly localized neurotransmitter, take the substance known as P. This neurotransmitter is a peptide released from the tips of sensory neurons, and it is important in the transmission of pain and inflammation. Although most neurotransmitters act in simply inhibiting or exciting a cell across a synapse (synaptic transmission), some act together in a more concerted way. These are called "neurotransmitter systems," and they include the dopamine system, the serotonin system, the norepinephrine system, and the acetylcholine (more generally, the cholinergic) system. The first two of these are probably most familiar to you: dopamine through its association with Oliver Sacks and Robin Williams, and serotonin because of its role in depression.

In the movie *Awakenings*, Robin Williams played Oliver Sacks as a young physician investigating the strange condition known as encephalitis lethargica—a syndrome with unknown origins but probably connected to immune reactions that leaves some of the infected in a sleeplike phase. The individuals with this syndrome contracted it right after World War I and remained in a sleeplike stupor for several decades. Sacks prescribed the administration of L-DOPA, a dopamine precursor, to these patients. Dramatic release from the "sleeping sickness" was observed, indicating that a lack of dopamine was involved. Release from the sleeping sickness is, however, almost always short-lived after treatment with L-DOPA. This is because the neurotransmitter dopamine is involved in a process called "neuromodulation." Unlike direct synaptic transmission, a neuromodulator is released and diffuses to bathe a large number of synapses. The bathing of several neurons from a single release event then induces an excitatory or inhibitory response by the nerve cells that are being bathed. What Sacks was able to induce by administering L-DOPA in the lethargica patients was neuromodulation that excited previously lethargic nerve cells.

Cleanup on Aisle Synapse

One important aspect of neurotransmitters and neuromodulators is that they continually need to be swept up and reused or discarded. If they simply remain stuck to the receptors on the postsynaptic neuron surface, or if a bunch of transmitters sit around in the synapse, then the subtle concentration changes of neurotransmitters needed to inhibit or excite a neuron are not detected by the postsynaptic cell. Imagine this: acetylcholine is released into the synapse of two interacting neurons by the presynaptic cell. The small transmitter molecule finds receptors on the membrane of the postsynaptic cell and interacts with them. An electrochemical response is induced in the postsynaptic cell, and an action potential is sent down the axon. If the acetylcholine is not cleaned up, swept away, or scrubbed off, then the synapse will subsequently be useless. This is because if another action potential makes its way to the presynaptic cell, signaling it to release more acetylcholine, the release will mean nothing to the postsynaptic cell because all of its receptors for acetylcholine will be clogged.

As it turns out, each neurotransmitter has a special way of dealing with the problem of resetting the synapse, much as we all have different ways of cleaning dirty floors. One cleaning strategy is to use a powerful cleaning agent such as Ajax that will break up the dirt on the floor. Another strategy is to get out the Dustbuster and vacuum. But probably the favorite is to sweep the dirt under the rug. These different cleaning strategies are actually pretty close to the way neurons clean up neurotransmitters. Some cells use the strong cleaning agent approach on acetylcholine. This transmitter is degraded in the synapse by an enzyme called acetylcholinesterase (the strong cleaning agent) that breaks the acetylcholine down into two inactive compounds. Serotonin, the other neuromodulator system we mentioned above, has a very interesting way of being scrubbed up. Serotonin is active in many physiological responses, but it is intimately connected to food and the digestive tract, where it is found in large amounts (up to 80 percent of all of the serotonin in your body is found in the gut). It is also, of course, found in the brain. There, serotonin is released like other neuromodulators, mostly from a localized area called the raphe nuclei, which lies in the brain stem. The cells in the raphe nuclei extend to other parts of the brain, and serotonin travels to these other regions via the axons of the raphe nuclei, which reach almost every part of the brain. Most serotonin receptors on postsynaptic neurons work by GCPR and second messengers (see chapter 2). To stop the response of the postsynaptic cell, the serotonin is "sucked" back up by the presynaptic cell, rather like vacuuming to clean up. This process is called "reuptake." Finally, dopamine simply diffuses away and collects in the kidneys, where it is degraded, rather like sweeping the neurotransmitter under the rug, the rug in this case being the kidney.

In the interactive exhibit we just described, the neural cells in the kid's brain experience synaptic transmission, neurotransmitter activity, neuromodulation, and the cleanup process, over and over again in a short time. The brain and the neurochemicals involved are acting fast and furious. The cleanup process is rapid and efficient, as the kid changes moods and physiology on the fly. And the kid can react with a range of emotions as a result of the activity of the neurotransmitters, the cleanup process, and the overall pattern of the action potentials that course through the brain. The choices made—whether the person interacting with the exhibit chooses (A) to be bad and get the cookie, or (B) to be good and go outside to play—illustrate another level of brain involvement. These are decision-making processes of the brain that are based in the prefrontal cortex (chapter 6).

The Good, the Bad, and the Ugliness of Neurotransmitters

Because each neurotransmitter has a specific pathway for getting cleaned up, different external forces can be used to target specific transmitters without interfering with other ones. Serotonin, for example, uses the reuptake process to clean up, and its effect can be modulated by controlling the reuptake process. Antidepressants are often based on specific disruption of the synapse cleanup process. In treating depression this disruption can be general or more targeted, as with the new selective serotonin reuptake inhibitors (SSRIs). Serotonin as familiar in humans also looks a lot like compounds that come from other organisms and like others that can be synthesized. These natural and synthetic compounds are called agonists, and they can interfere with the receptors on the postsynaptic cell. When delivered to the brain, they "halt" the cleanup process and short-circuit the synapse.

One effective way to upset the serotonin balance in the synapse is to get the serotonin vacuoles to release serotonin in inordinate amounts. This is how the empathogen-entactogen MDMA (X, or Ecstasy) works. The flooding of the synapse with serotonin short-circuits the synapse, and the various highs experienced by users of Ecstasy ensue. The "classical hallucinogens," drugs like psilocybin, lysergic acid diethylamide (LSD), and mescaline, also act as serotonin agonists and affect receptors in the cortex of the brain. These classical hallucinogens are potent agonists, and they can modulate the interaction of serotonin and glutamate in the prefrontal cortex. In a nontripping brain, serotonin modulates the release of glutamate and ensures that a manageable amount of glutamate is released in the cortex. But when too much glutamate is released, the brain starts firing in strange ways, and "tripping" results.

The unnatural interaction of the serotonin and glutamate systems in psychedelics is an example of two neurotransmitter systems clashing. Another class of such interactions occurs in addiction. In this case neurotransmitters called endorphins are released by the pituitary gland and enter the bloodstream, and the hypothalamus releases endorphins to the brain. The endorphins have a high affinity for a class of cell neuroreceptor molecules called opioid receptors. These receptors come in several flavors, but the one that forms part of the interaction of different neuromodulator systems is called the "mu opioid." Endorphins have a high affinity for this receptor, which is usually a presynaptic neuron receptor that affects neurotransmitter release. They specifically inhibit the release of GABA and disinhibit release of dopamine. All is well when the system is balanced by our own physiology, but when outside molecules are introduced to the brain that act as agonists of endorphin and bind to the opioid receptors, bad things happen. Externally ingested, snorted, or injected opioid receptor agonists will cause an unnatural release of dopamine.

Dopamine is the key here: it helps our brain to modulate our expectancies of pleasure and reward. How did dopamine get such an important job? Dr. Nora Volkow, director of the NIH National Institute on Drug Abuse, explains that organisms have always needed some form of reinforcement to recognize that good things were happening and to encourage them to pursue those good things. Dopamine thus evolved as the neurotransmitter that implements the conditioning of organisms to do things that are needed for survival, via giving a pleasure sensation to the brain. The pleasure sensation associated with a specific act like eating or having sex motivates us to do it again and again. And when exogenous sources of endorphins are introduced to the system, and dopamine is unnaturally released to the inner regions of the brain ("the limbic system") in large amounts (up to ten times more than usual), the brain begins to say to itself, "If I do this drug, then I will really get a big payoff in the pleasure department."

The big rise in dopamine levels caused by the introduction of opioid receptor agonists like cocaine and heroin will be misinterpreted by the brain as being normal. The need for the brain to maintain an up-regulated "normal" level of dopamine will override everything; and since the drug is the only thing that can meet the brain's expectation, it will be used over and over again. Essentially, what happens in addiction is that the consistently high levels of dopamine in the brains of users create physical and lasting changes to the brain. Neurons become as it were numb to the agonists and resistant to being affected by the drug. This in turn leads to loss of receptors in the synapse. The system then gets shorted out, much as in the serotonin system we discussed above, but with terribly opposite repercussions. Actually, "shorted out" is not the best way to put it. Adam Kepecs of Cold Spring Harbor Laboratories, who studies the neurobiology of decision-making, calls it a "neural hijacking" and explains it this way:

Let's say you're happy about a great chocolate ice cream. Over time you learn to expect that the chocolate ice cream is really great and you have no more dopamine released in expectation of that when you receive it. Whereas, if you take an addictive drug, you can never learn to expect it, because the drug itself will release an extra kick of dopamine. And when that happens, the value of that drug keeps increasing because now you're learning that "Wow my expectations were violated, therefore this must be much more valuable than what I thought before." So basically what ends up happening: the dopamine system gets hijacked by these drugs."

One of the critical aspects of addiction is that, as the hijacking is being done and the user is exposed more and more to the drug, the brain quickly learns that the drug is needed; and memory, specifically a kind of memory called conditioning, will reinforce the desire for the drug. The hijacked brain is now at a complete disadvantage, with pleasure reinforcement and conditional memory working against it. Addiction ensues when the drive to consume the agent that increases the dopamine level becomes uncontrollable. Emotions and memory are key here; but they are also part of the story of how we interpret the outer world, so we turn now to these important aspects of the brain.

6 EMOTIONS AND MEMORY

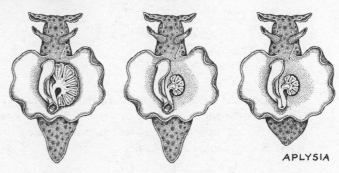

APLYSIA

Emotions are an important innovation in the history of life on earth. The parts of the brain sometimes assigned to the limbic system are varied in their function, but in general they are involved in the processing and expression of emotions. These structures include the brain regions such as the amygdala, the hippocampus, the cingulate cortex, and the hypothalamus, which all form a kind of loop in the inner region of the brain. Because lizards and birds have limited versions of these regions, it is not surprising that major differences in behavior between them and mammals lies in the emotional components. Lizards are, so to speak, emotionless, reflexively responsive creatures. Not that this is necessarily any handicap or a sign of being "less evolved." Many organisms get by quite efficiently without emotions because they have no specific need for them. Single-celled organisms, very simple animals, and even some very complex animals do not have emotions. To suggest that a hydra, a fruit fly, or even a lizard feels happiness or sadness would be to overrate their nervous systems. They simply do not have the neural basis for emoting. Conversely, understanding our limbic systems becomes an important aspect of how we understand our emotions.

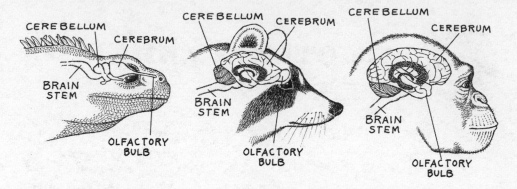

Comparison of the limbic system in three vertebrates, lizard (left), raccoon (middle), and primate (right). The limbic system is colored in dark. The cerebellum, brain stem, cerebrum, and optic bulb are labeled as reference points. Note the lack of the dark-colored region in the lizard.

Once more, Aristotle is a good place to start. Although he didn't think the brain had much to do with thought or consciousness, he did state clearly that emotions such as laughter were an exclusive trait of humans. Next, skip two thousand years or so to Paul Broca, that great nineteenth-century student of the brain who explored many aspects of its anatomy and who recognized the importance of the part that he called the limbic lobe. Although Broca focused on the putative olfactory functions of this part of the brain, he did recognize its importance in emotive behavior (in a very visceral form that he called l'*homme brutale*). Perhaps, then, a better connection of the brain to the emotions came from Charles Darwin. No one better articulated the link between animal and human emotions better than he did in *The Expression of the Emotions in Man and Animals* (1872). In this wonderful and innovative book, Darwin drew attention to extensive similarities between human beings and other mammals in facial expressions, reactions, and postures. The brilliance of Darwin's analysis lies in his suggestion that these similarities were inborn, the results of common ancestry and (perhaps more important) the products of natural selection. Why else would a dog bristle and a man puff out his chest when confronted? Why do we bare our teeth when angry or fearful and animals do the same?

Darwin recognized that some of our emotions fell into in this primal category but also perceived that others seem more advanced. Still, almost all of the human emotions he focused on had the same characteristics as those exhibited by mammals. And although Darwin did comment on lizards and birds in this amazing book, he focused most of his evolutionary explanations on the mammal-human connection.

Darwin commented largely on such emotions as sadness, anger, happiness, and fear. He noticed that actions and reactions of many mammals resembled the ways in which humans emote, as in laughing. Smiling, he suggested, "may be said to be the first stage in the development of a laugh." Laughter could have evolved, he thought, because "the same muscles are brought into slight play whenever any cause excites in us a feeling which, if stronger, would have led to laughter; and the result is a smile." More recently, some animal behaviorists have suggested that laughter is more widespread among mammals than previously thought, even more widespread than Darwin suspected. The psychologist Jaak Panskepp, for one, has studied rat laughter. He even thinks that rat humor might be based on slapstick comedy. This might seem improbable, but bear in mind that we can't hear rats laugh very well because they laugh at a frequency much higher than our range of hearing (rats chirp at approximately 50,000 Hz, and we normally hear between 25 Hz and 20,000 Hz). What's more, you need either to tickle the rats to make them laugh or catch them when they are playing. Panskepp and his colleagues are convinced that the rats they work with are laughing (see for yourself at http://graphics8.nytimes.com /packages/video/science/rat.mov).

Panskepp is careful when he tries to make the leap to human laughing, but here we encounter the metaphor problem again. When looking for explanations for phenomena in nature, we have to be careful not to fall into the storytelling trap (chapter 1). If rat chirping is the "same" as human laughter, then we would expect to see it in many more mammals than we currently do. It is more likely, then, that rat chirping is a convergence on human laughter. Still, Panskepp points out, even if it is just a metaphor, we have a lot to learn from rat laughter, at both the behavioral and genetic level. Who knows? Maybe their jokes are really, really funny.

I'm a Little *Verklempt* (Mike Myers)

Emotions are such a slippery subject that we wondered whether we should overlook them in this book. Jaak Panskepp addresses the same dilemma in *Affective Neuroscience: The Foundations of Human and Animal Emotions*, when he asks, "Do neuroscientists need to understand emotions to understand the brain?" And he resoundingly answers in the affirmative: "Newfound information about the brain, with the many anatomical, neurochemical, and neurophysiological *homologies* that exist across all mammalian species, has the potential to render such neuro-mental processes as emotional feelings measurable, manipulable, and hence scientifically real." To Panskepp, what was once an alchemic endeavor now has a periodic table to make it more scientifically sound. In his view, emotions are essential to our understanding of the brain, and moreover, emotions are approachable. But they are still pretty elusive, because in some ways their recognition and naming are somewhat subjective and unscientific. For instance, Aristotle recognized twelve emotions; the philosopher Baruch Spinoza categorized about forty-eight of them; and the psychologist B. F. Skinner suggested that we have no emotions at all.

More recently, psychologists like Panskepp, using clever experiments with both animals and humans with brain lesions, have pinpointed five emotions: happiness, sadness, anger, fear, and disgust. And they have undertaken some difficult studies in the attempt to understand them. They have involved the tried-and-true approach of finding individuals with specific lesions in the brain that evoke a defined psychological defect or alteration. One of the better-known examples of this approach involves the examination of several people with damage to the amygdala on both sides of their brain. The amygdala are small, bilaterally arranged, deep brain structures sometimes assigned to the limbic system and sometimes to the basal ganglia. When shown pictures of human faces with facial expressions for different emotions, individuals with bilateral amygdala damage will not react to faces whose expressions show fear, which demonstrates that the basic understanding of fear can be altered by a loss in function of a particular region of the brain. By looking at a range of individuals who show different psychological alterations when exposed to the same facial expression, we can "atomize," or tease apart, the variety of emotional responses.

Other researchers, such as the neurobiologist Antonio Damasio, like to discriminate between emotion and what they suggest is a higher level of behavior called "feeling." Damasio suggests that emotions and feelings are tightly bound to each other but that one (emotion) is a public expression of the brain and the other (feeling) is a private thing. What's more, emotions come before feelings: "We have emotion first and feelings after because evolution came up with emotions first and feelings later." To Damasio, emotions are the result of simple reactions that "promote survival of an organism and this could easily prevail in evolution." In weighing such suggestions, we must consider that, as mammalian evolution proceeded, the cortex grew to be a substantial part of the mammalian brain and that, in primates and especially in ourselves, its "prefrontal" portion enlarged greatly over evolutionary time. This fact is vital because the prefrontal cortex ensures that we are not slaves to our emotions, allows us to process the emotional responses of our inner brain, and enables us to use other decision-making mechanisms to modify responses to external stimuli in ways that are sometimes counter to our emotions.

Arguments such as Damasio's support Darwin's suggestion that evolution is the best explanation for why we have emotions. Darwin was initially puzzled why our hair (or what is left of it) stands up when we humans are scared or threatened. He ultimately decided that this response had evolved in lower mammals as an adaptive defensive response and persisted in the descendants of those ancestors. He made the same conclusion for other emotions he observed in both humans and other mammals: we retain them because they were beneficial to an ancestor. In discussing emotions in *Looking for Spinoza,* Damasio creates a hierarchy of organismal traits that preclude and include emotions in organisms. He points out that the various levels of the hierarchy are nested traits. The most primitive is nested within the next level, which in turn is nested in the next higher level. For instance, he recognizes that organisms have very primitive communication systems, much like the systems we discussed in chapter 2.

These include the basic reflexes and the immune response. The next level includes the basic pain and pleasure behaviors. The penultimate layer, before reaching emotions and feelings, is where drives reside.

Damasio then suggests that there are three basic kinds of emotions: background, primary, and social. He proposes that these three basic kinds of emotions are also nested—that is, background emotions are a specific kind of primary emotions, and primary emotions are a specific kind of social emotions. This nesting implies an evolutionary pattern with more

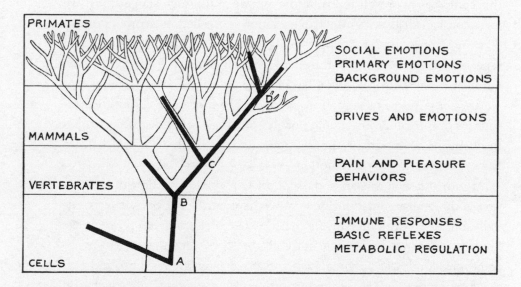

Antonio Damasio's tree of feelings redrawn to show how a phylogenetic view might work. The star at the bottom represents the ancestor of all life on earth. The open circles and letters represent the ancestors that acquire the traits discussed by Damasio and shown on the right. Although it is difficult to determine what organisms these ancestors gave rise to, we can estimate that ancestor A is the ancestor of Bacteria, Archaea, and Eukarya. B might be the ancestor of the Metazoa. C might be the ancestor of vertebrates and D the ancestor of mammals. This does not mean that other organisms do not have the traits listed on the right. In fact, if these other organisms, like plants and Bacteria, have these traits or evolve to have them in the future, then the traits in these other organisms are or will be convergences.

primitive organisms showing the least inclusive category, background emotions. Organisms that have nervous systems also have the primary emotions, and organisms with highly developed nervous systems have the social emotions. In fact, Damasio uses a tree as an explanatory device to show this hierarchy. Unfortunately, his tree branches all over the place. With apologies to Damasio, we have redrawn this tree to look more like an evolutionary tree to demonstrate that nesting can also be looked at as an evolutionary process. This way of looking at emotions is both interesting and illuminating: because some of these evolutionary significant steps in emotions are correlated with changes in brain structure, it tells us something about how emotions actually work at the molecular and physiological levels. But how does outside stimulation result in those little *verklempt* moments we all have?

So That We Might Have Roses in December

One of the big mysteries of the brain is how organisms store and interpret past experience. And important aspects of this phenomenon bear on our consciousness as humans. Our ability to think easily in the context of time, to anticipate the future, and to synthesize the past are all important aspects of the uniqueness of our brains when compared to other brains on the planet. But other organisms can store and use memories. The simple sea slug *Aplysia*, with its twenty thousand neural cells, has been widely used to understand the basic neurochemistry of memory. Yet the term "memory" is broad, and the various kinds of memory that have been characterized in the neurobiology literature can be hard to keep track of. The situation is particularly tough because any possible way to subcategorize memory usually turns out to be an oversimplification.

Still, some systems of classifying different aspects of memory have merit. In this book we use a "two-dimensional" way of looking at memory. In one dimension, the length of the memory is considered—that is, long-term versus short-term memory. Short-term memory is the capture by our sensing systems of fleeting sensory experience, and it involves two aspects of the immediate treatment of outside experiences by our brain. First comes sensory memory. As we discussed in chapter 4, this is the kind of neural signaling that is delivered to the brain from the six senses. A sensory memory is a direct copy of the experience, and it lasts a very short time. Any sensory memory that the brain deems useful is transferred to short-term memory. Short-term memory is then conveyed to other regions of the brain. Think of both sensory and short-term memory as filtering devices that essentially control the storage of sensory input. If the sensory memory is not passed on to short-term memory or short-term memory is not fortified or explicitly repeated, then it won't be passed on to long-term memory. Anything that isn't too terribly important to the survival of the organism is lost from short-term memory and is simply not passed on for further integration into the long-term memory pathways.

Our short-term memories are limited. If you couldn't remember what was being said in the first part of this sentence for more than a second, then the end of the sentence wouldn't make sense. The normal extent of short-term memory is about seven bits, or items, that can be remembered. The smaller the memory task, the more likely it is that it will stick. So if you are confronted with the number 369201144830, there is only a slight chance that you will remember that number by looking at it for ten seconds. Try it. But if you spent that same ten seconds splitting the number into four groups of three (369 201 144 830) and quickly associated those four smaller numbers with easily remembered episodes in your life (multiples of three, your homeroom in high school, a gross, and the time of the commuter train's arrival each morning), then the task is doable. This process is called "chunking," and it is also used by the best chess players in memorizing or recalling chess moves and strategies. If you are playing someone who can chunk and you can't, then your best strategy is to quit. According to our colleague Alex deVoogt at the American Museum of Natural History, chunking is also used in African and Southeast Asian board games like mancala and boa. These games require rapid computational and memory skills that are best played using a chunking strategy. Some mancala players apparently know exactly how a game will play out after only a few moves, and this is because of memory from previous scenarios of play. Likewise, if you are playing mancala with someone who can chunk, don't make any bets.

In the other dimension of memory, we can look at this ability as a function of flexibility or reflexivity—that is, whether the memory is declarative (highly flexible) or nondeclarative (little or no flexibility). Nondeclarative memory is responsible for repetitive actions like riding a bike or the motions you use in brushing your teeth. Declarative memory is, in contrast, the kind of memory we used in writing this book when we drew on our knowledge of evolution and biology to put words on the page. One interesting way to keep these kinds of memory straight is to think of declarative memory as what you might use to remember why you brush your teeth and of nondeclarative memory as what you use to control the brush movements when actually brushing your teeth.

Declarative memory (also sometimes called explicit memory) is a complex process that can be subdivided into episodic and semantic memory. The difference between these two kinds of declarative memory is the degree of personal context each has. Semantic memories are impersonal and are disconnected from the autobiographical past of the individual having them. Episodic memories, in contrast, are highly personal and autobiographical. A semantic memory might be that you know 9/11/2001 was a date on a calendar in September 2001 and that if that date were a Tuesday, then Wednesday would follow. An episodic memory would be a memory of where you were and what you were doing on 9/11—for example, that it was perfectly sunny in New York City and you were sitting in your office in the American Museum of Natural History (more on that later). As for where in the brain these kinds of memories reside, there is some controversy. Declarative memory is thought to be modulated by the hippocampus, that important structure of the "limbic" system. Semantic memory is thought by some neuroscientists to reside in the same area (the hippocampus) and by others to be more of a prefrontal cortex function.

According to Katharina Henke, nondeclarative memory (also loosely known as implicit memory) can be subdivided into four distinct categories based on the location in the brain of the memory function or storage. The major category is "procedural memory," which is important in the memory of skills and habitual behavior (like the strokes you use to brush your teeth). This category of memory resides in the basal ganglia. The second kind of such memory is "priming," which can be mapped to the neocortex. "Classical conditioning" is a kind of nondeclarative memory that functions in the amygdala and cerebellum, whereas "habituation" and "sensitization" are localized to the reflex pathways of the central and peripheral nervous system. We will discuss some of these forms of memory shortly, in the context of molecular control.

Not as Pretty as Heidi Klum but Nevertheless a Darn Good Model

Until recently, what we knew about memory came from two major kinds of scientific approach: studies of nonhuman brains that didn't mind your telling them with whom to mate (the model organism approach) and studies of cracked-open human brains (the clinico-anatomical approach). In the first approach, there is an art to choosing a model organism for studying memory or indeed for any kind of biological study. One should choose not only something that won't bite you but also a model that will have certain qualities that make it worth the trouble. Usually the researchers who choose the models are consummate naturalists who know something about the organism that they select that few others know. And through cleverly designed experiments, they take advantage of the traits of the model organism to uncover the mysteries of the genes, neural connections, and cells. An important nonhuman animal model for memory research is actually not the familiar *Drosophila*, the nematode, or even the mouse but the lowly sea slug *Aplysia*. Actually "lowly" is a misnomer, because sea slugs are every bit as related to us as is a nematode or a fly. They are all protostomes, and hence they share the same common ancestor with us. *Aplysia* was chosen as a model organism for studying memory by Eric Kandel of Columbia University, who chose this animal partly because it was easy to grow in the lab but mainly because it had nice big juicy nerve cells (up to 1 mm in diameter) and very few of them (twenty thousand total).

But Kandel also selected *Aplysia* because of its interesting behavior when prodded. This animal has what is affectionately called a "gill and siphon withdrawal reflex." If one touches the siphon of the slug, this causes the

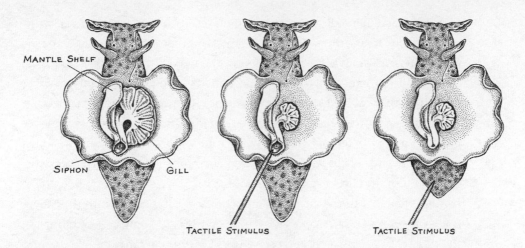

Aplysia, the sea slug, and its reflex system, which Eric Kandel and colleagues exploited to study memory.

siphon to contract and the gill to withdraw, like a worm crawling back into a hole, to protect these important protruding organs of the slug. It is an unspectacular response to a simple stimulus. With this response in mind, Kandel did what we call a classic "give it to the undergrad intern" experiment. When the slug extends its siphon again to feed, if the siphon is touched slightly once more, the gill and siphon will contract again, but not as rapidly and as extensively as the first time. Each time the siphon is touched, the degree of response by the slug is recorded. If this approach is repeated seventy to eighty times, with three-minute intervals between the stimulations, then by the seventy-ninth touch the slug will barely contract at all. The slug has "habituated" to the stimulation of the siphon. Classical habituation will erode after allowing the slug to rest for a while; indeed, if the undergrad goes out for lunch for two hours and comes back and touches the slug's siphon (eightieth replicate), the response will be as extreme as when the experiment started. This form of habituation is similar to habituation in vertebrates and constitutes a form of nondeclarative memory. So Kandel reasoned that *Aplysia* would be a good model for understanding the neurochemical basis of memory (or at least of nondeclarative memory).

Kandel and his colleagues at Columbia used the gill and siphon withdrawal reflex to ask three basic questions about memory: (1) What is the locus of memory storage for habituation in the nervous system? (2) Do plastic changes in elementary synaptic connections contribute to memory storage? and (3) If so, what are the cellular mechanisms for memory storage? Because *Aplysia* was a relatively new model system, getting at these three questions required a lot of procedural grind. Yet, with great application, Kandel and his colleagues were able to work out the neural circuitry of the gill siphon withdrawal reflex, to develop a method of recording the firing (action potentials) of single neurons in the circuit, and to correlate habituation with the movement of vesicles near synapses. The next step was to apply these findings to the three questions Kandel and colleagues had set out to answer.

The first question, it turned out, had a relatively simple but elegant answer. But before getting to it, a paradox of the nervous system had to be solved. This paradox resulted from the observation that, while neural connections in the sea slug nervous system are predetermined developmentally for all sea slugs, there is nonetheless some plasticity with respect to how this hard-wired system reacts to different stimuli. Even more important, the same system can reset itself after habituation. If the circuitry really is invariable, then you have to solve the paradox before you can explain how different memories can be stored in the same circuitry. Kandel solved it by showing that even though the circuitry is invariable, the strength of the neural connections varies from time point to time point. This gave him the answer to the first two questions: the locus of memory storage is indeed in the synapses, and, yes, plastic changes in synaptic strength are at the heart of how nondeclarative memory is stored in the nervous system. The third question was harder to answer, but because Kandel could view changes in the synaptic morphology of habituated, recovered, and naive synapses, he could correlate the release or nonrelease of vesicles at synapses with the state of the animal's habituation. This showed that habituation significantly lessened vesicle contact with the synapse. In other words, the synapses were rendered less "hot" by habituation.

To put this all in context, these experiments allowed Kandel and colleagues to demonstrate a fundamental principle of the nervous system—

one that connected all kinds of previous work done on the nervous system. They even referred back to Santiago Ramón y Cajal's early suggestion (see chapter 3) that synapses are not fixed but that their strength can be altered through time by experiences and repetition. More precisely, they changed the way we thought about memory. Before Kandel's work, memory was thought to be stored in the brain much as it is in computers: simply assigned to a location, perhaps even to a single neuron or a clump of them. But the *Aplysia* experiments clearly showed how much more complex than that memory storage is and that the *strength* of the neural synapses is involved in how memory works. And Kandel correctly deduced that this had something to do with the chemical workings of neurons and their synapses.

So he and his colleagues took the *Aplysia* model a step further. Notice that so far Kandel had been able to deduce these seminal tenets about memory without breaking into a cell. All observations up to this point had been made either by tracing neural connections, by measuring action potentials, or by manipulating the behavior of the sea slug. Because the changes in the vesicles of neurons appeared to be intricately involved in the plasticity of the synapses, Kandel decided to go into the neurons themselves and examine what was going on chemically in the vesicles. To get at the chemical nature of memory, Kandel used two other easily manipulated aspects of *Aplysia*'s behavior. Habituation is a simple response, so Kandel decided to use the more complex nondeclarative responses called sensitization and conditioning.

Sensitization is to some extent the flip side of habituation. Unfortunately for the sea slug, inducing sensitization requires applying a slight electrical shock to the tail, which causes an extreme siphon and gill withdrawal response. After the administration of a single shock to the tail, *Aplysia* will retain the memory and will withdraw its siphon and gill automatically and strongly. After several minutes, though, the slug will reset its memory of the shock and discontinue withdrawing the gill and siphon strongly upon shocking. If, however, the animal is shocked four or five times consecutively on the tail, it will remain sensitized for two or more days, during which its response to a shock will be extreme. With further training, even longer periods of sensitization can be induced in *Aplysia*. Since the tail is involved, Kandel had to do yet more procedural work to figure out the neural circuitry of this part of the sea slug's body, reasoning that, if habituation resulted in a decrease in synaptic strength as a result of vesicles moving away from the synapse, the sensitization would produce an increase in synaptic strength and a correlated increase in vesicle contact with the synapse. Note that, with habituation, the upshot was that the organism ignored the stimulation after a while because, time after time, nothing bad had happened. In sharp contrast, in sensitization the neural system amplified the response to continued shock because the shocks were so obnoxious. But what would happen when an experiment got the animal to correlate something bad, like a shock, with a harmless stimulation like tickling the siphon? This is called conditioning, described by Kandel and Larry Squire as follows: "The predictive rules that characterize classical conditioning mirror the cause and effect rules that govern the external, physical world. It thus seems plausible that animal brains have evolved neuronal mechanisms designed to recognize events that predictively occur together and to distinguish from events that are not related to each other." Amazingly, Kandel and his colleagues were able to induce conditioning in *Aplysia*, an extremely simple animal, just by using the siphon and gill retraction. As a result, with the tools of sensitization, habituation, and conditioning in hand and *Aplysia* as a "work-slug," scientists guided by Kandel's groundbreaking work were able to examine important aspects of the neurochemistry of memory.

Phineas Gage, Meet Henry Molaison (H.M.)

The oldest and crudest way to tackle the problem of memory is to look for someone with a behavioral disorder or a brain injury, characterize the person's behavior, then wait for him or her to have a brain operation or, worse, die. Then you crack open the skull and examine the brain for anomalies or lesions. If the brain lesion is localizable while the person is alive both he or she and you are in luck, as death will not be required for some inference to be made. This "clinico-anatomical" approach has been used for 150 years and was initiated by Jean-Martin Charcot, renowned as the father of neurobiology. Charcot was a professor of neurology in Paris, and after an illustrious career had a disease named after him (Charcot-

Marie-Tooth disease, which has nothing to do with teeth). He was also the first to formally describe and name Lou Gehrig's disease (he named it *sclerose en plaques*, decades before Gehrig was born), had an island named after him in Antarctica (Charcot Island), and was known as the "Napoleon of the neuroses."

Although the epicenter of the clinico-anatomical approach was 1870s Paris, perhaps the most famous such case had occurred twenty years earlier in New England. It resulted from a gruesome incident involving a gentleman named Phineas Gage. Gage was a railroad construction worker who set dynamite charges for blasting rock. The procedure was rather crude back then, and in the process of compacting dynamite into a hole with a "tamping iron," Gage experienced a terrible accident. Tamping irons were almost two meters long, with a pointed end, and while tamping Gage somehow produced a spark and an explosion that blew the tamping iron out of the hole in the rock and straight up and through his head. It entered his left cheek just behind his left eye and tore through the prefrontal cortex of his brain and out the top middle of his head, landing several yards behind him. Needless to say, this had a profound effect on Gage, but it wasn't death. He survived the incident intact (except for the loss of a good chunk of his prefrontal cortex), saw a doctor, walked home, and came to work the next day. To doctors who examined him it was obvious that the front part of his cortex, that lump of brain behind the forehead, had been terribly damaged; and yet he could walk, talk, and remember things.

Still, formerly a temperate, religious, hardworking, and helpful man, Phineas Gage turned into a drunkard and a swearing, belligerent sloth. As some of his friends said, he was "no longer Gage." And he couldn't help it. We now know that the prefrontal cortex is the seat of reason and personality. Whenever it is damaged, the personality of the person involved changes drastically. Many examples of the use of disease-related or genetic brain lesions have helped in uncovering brain function, although most of them are less spectacular than the case of Phineas Gage. The most famous of them concern the eponymous regions of the brain involved in language perception and production that were discovered by the nineteenth-century physicians Paul Broca and Carl Wernicke.

Since the days of Broca and Wernicke, numerous examples of patients with specific brain disorders or genetic syndromes have been used to pin down the location of memory in the brain. A convention used by the doctors and scientists who study such patients is to identify them by their initials. Perhaps the most famous of the memory studies using this approach is the case of H.M., who had severe epilepsy. When H.M. died, he lost his anonymity, so it is now known that his full name was Henry Gustav Molaison. To relieve H.M.'s seizures, a Connecticut surgeon named William Scoville removed the area of his brain that was thought to be causing the seizures—namely, the entire hippocampus and medial temporal lobe. The surgery brought the seizures under control, but afterward H.M. was unable to store memories. He could recall memories from before the surgery, but his current life consisted of two-minute vignettes that were not stored in his long-term memory. Because of his unique psychological state, H.M. became the subject of many psychological tests conducted by Brenda Milner, a Canadian psychologist. Milner came to several important conclusions. The first and most obvious was that obtaining some kinds of new memories was a function of the regions of the brain now missing in H.M. Second, Milner deduced that the medial temporal lobe and hippocampus, the removed regions, were not the ultimate point of storage of long-term memories, because H.M. was able to recall memories from before the surgery. Milner also deduced that these regions were not necessary for immediate short-term memory, because H.M. could retain small amounts of information for a few minutes. Instead, the medial temporal lobe and the hippocampus were apparently responsible for some aspect of transmitting short-term memory into long-term memory. Milner's final big conclusion was based on a common psychological test known as the star-tracing test. In this test, the subject is asked to trace a star, as in the adjacent figure. Sounds easy, but when the only image you are allowed to see is the star in a mirror, it becomes quite difficult. The first time most people try it they make many mistakes, but after several attempts they get better and better. This is because the person doing the test starts to store the memory of how to do the tracing in his or her procedural memory (which, as you'll recall, is a form of nondeclarative memory). H.M. was given this test, and over a period of three days he went from making twenty to thirty errors in the star tracing to only one or two errors. But when he was asked before each test whether he had ever taken the star test before, he would invariably answer no, that he had never even heard of such a test. So it was evident that H.M. could store procedural memories even though he could not store long-term declarative memories—indicating, of course, that that the part of the brain removed by the surgery had not affected procedural memory.

Other aspects of memory have been studied using patients with naturally occurring brain lesions. And an exciting recent development is that nowadays neuroimaging approaches like magnetic resonance imaging, positron-emission tomography, and others can be used to localize the lesions with precision and to correlate those lesions with certain functions. Notger Muller and Robert Knight used a survey of several people with brain injuries to study a type of memory called working memory. These patients all had specific brain lesions caused by head trauma, stroke, or invasive surgery. The researchers examined the abilities of the patients to use and manipulate working memory; and using MRI approaches they were able to scrutinize the structures of their brains without having to

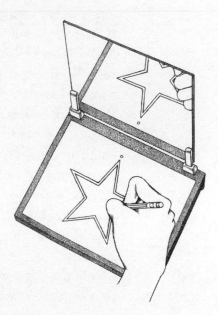

Star tracing diagram used for the mirror tracing test that H.M. was given to examine procedural and declarative memory.

crack open any skulls. They concluded that lesions involving the dorsal and ventral prefrontal cortex significantly impact working memory. Even more important, Muller and Knight emphasized the importance of examining behavior in conjunction with localization of lesions in the brain. Patients can be instructed to perform specific mental tasks while undergoing imaging, and the resulting brain activity can be closely examined. As imaging techniques improve, the payoff promises to be even greater; Muller and Knight point out that novel approaches such as diffusion tensor imaging and a related approach called diffusion sensor imaging will add greatly to the understanding of brain lesions and function, not only in memory but in other areas of cognitive functioning as well.

LTP? OMG CRS

When we look at long-term declarative memory from the perspective of the neuron, the work of Eric Kandel and others on short-term memory has huge relevance, even though the types of nondeclarative memory that Kandel and others examined back in the twentieth century are somewhat different from declarative memory. The trick for understanding why this should be so lies in understanding how a neuron can hold a signal for a long period of time. This should be simple to visualize, because the signal is clearly somehow responsible for the memory, and if it weakens or gets lost, the memory should get lost, too. How the two are united from a neurochemical standpoint is a fascinating story that starts with Donald Hebb, a neuroscientist who first worked with the famous neurosurgeon Wilder Penfield in Canada and then on his own for a good part of his career. In the late 1940s Hebb questioned the accepted convention that complex circuitry in the brain was needed for learning and memory and suggested instead that strengthening synapses is the reason learning occurs. He felt that the process of storing memories *must* involve the neurochemistry of the synapse. This is pretty much the same idea Cajal had and that Kandel later expanded by suggesting that synapse strength lies at the heart of memory.

To understand what it is that changes the synapse, we need to go back to how synapses work. It makes sense that any differences you might find in memory synapses from the steady-state workings of synapses in general should be unique to memory. In synapses going about their daily business, activity in the presynaptic cell will induce a firing of the postsynaptic cell, and this firing will in turn strengthen the synapse. Starting from this proposition provides a completely different way of looking at learning than does the lens of complex circuits. But we have to know how the synapse holds the signal and/or increases it. Long-term depression (LTD) and long-term potentiation (LTP) provide the answers. These important neural processes were first demonstrated in the hippocampus and related structures of the rabbit and later in the cerebral cortex, cerebellum, amygdala, and many other structures. LTP is perhaps the better-understood mechanism of the two and involves the interaction of neurotransmitters, receptors, and the synaptic membranes of communicating nerve cells. LTP was discovered in 1973 as an artificial response that rabbit hippocampus cells made to electrical stimulus.

When glutamate is released into a normal neural synapse (chapter 3), it will bind to specific receptor proteins on the surface of the postsynaptic cell membrane. The glutamate binds to two kinds of receptors: NMDA receptors and non-NMDA receptors. The glutamate binding to NMDA receptors causes the gated channel to become clogged with magnesium ions. The glutamate binding to non-NMDA receptors causes those channels to open and potassium and sodium ions to flow into and out of the postsynaptic cell. The postsynaptic cell's electrochemical milieu is thereby altered, and an action potential ensues in the postsynaptic cell. LTP involves different mechanisms and cellular changes that keep a neuron in a specific neurochemical state for a long time. LTP has two phases, labeled early and late. Early LTP is induced by rapid "high-frequency" firing of the presynaptic cell. This high-frequency firing results in the release of large amounts of glutamate and reduces the membrane potential of the postsynaptic cell. When this happens, the Mg+ ion that is lodged in the NMDA receptor jumps out into the synapse. The NMDA receptor is now unplugged, as it were, and calcium, sodium, and potassium ions start to move around. The upshot of this unplugging is that the calcium ion concentration on the inside of the postsynaptic cell rises to a critical concentration and causes a cascade of chemical reactions involving protein kinases in the postsynaptic neuron. These kinases are enzymes that add phosphates to other molecules, whether they like it or not, in a process called phosphorylation. This changes the conformation of the proteins that get the phosphates, and in some cases it activates, and in others shuts down, the activity of the target protein. Hence this cascade of reactions, caused by the initial unplugging of NMDA receptors, triggers a series of switches on the proteins of postsynaptic cells and changes the potential of the postsynaptic cell.

So far so good, but how is LTP maintained in the postsynaptic cell? The maintenance of LTP requires that another membrane receptor get involved. This receptor is called an AMPA receptor, and its production by the neural cell is controlled by one of the protein kinases we just discussed. With the AMPA receptor now sitting in the postsynaptic cell membrane, the calcium ion concentration that is critical for LTP can be maintained for a long time. Also, the non-NMDA receptors embedded in the postsynaptic membrane are acted upon by the kinases that have been switched on by the initial stimulus; after being phosphorylated, these non-NMDA receptors are more sensitive to the effects of glutamate in the synapse. The synapse has changed and is now much more efficient at managing the action potentials when it is charged with neural tasks. LTP has three major features that fit the bill quite nicely for neurobiologists interested in explaining memory. The first is "guilt by association." LTP occurs in each of two important neural pathways in the brain. Both pathways occur in the hippocampus and involve regions of the brain important in processing the information from our sense systems. It makes sense that these regions should be areas where a memory process is active. Next, LTP is fast and can increase the ability of the synapse to fire more efficiently and with precise timing—a really important aspect since we need to recall memories quickly and efficiently. Third, once LTP happens in a cell, it can be maintained by the neuron for long periods of time. This final aspect of LTP is especially important because, after all, memories are "pressed between the pages of our minds" and need to last a long time.

So we are pretty much there, right? We now have a molecular cellular mechanism that can alter the strength of synapses for long periods of time, so this must be it. Not so fast. As Squire and Kandel observe: "Simply because LTP has features in common with an ideal memory process does not prove that it is the mechanism used for memory storage in life." They say this because LTP was discovered in an artificial experimental system; hence, it is important to tie LTP to real-world long-term memory. Two major kinds of experiments have addressed the validity of LTP in long-term declarative memory. The first approach involves the use of swimming rodents; the second, the use of rodents with wires in their brains. Basically, the machinery of LTP is disrupted by genetic or chemical manipulation of the rodent. The altered rodent is then placed into a circular pool with a submerged platform that the rodent can't see and doesn't know about. Spatial indicators are placed on the walls of the room with the pool, and the rodent navigates the pool until it finds the platform and takes a rest. A normal rodent easily remembers where the platform is by orienting to the spatial signs on the walls of the experimental room, and finding the platform and correlating spatial location use its declarative memory and hippocampus.

In another experiment, using mice, the platform is marked with a tiny flag. The rodent finds the platform and correlates the flag with the platform. This activity uses nondeclarative memory and is a nonspatial process in the brain. Researchers used this approach (known as a Morris water maze) to see what happens when you alter the NMDA receptors needed for LTP to be established. When NMDA receptors are blocked by a chemical injected into the rodent, the rodents retain the nonspatial ability to find the platform, but their declarative memory is confused, and they fail miserably at the task. The results of this experiment suggest that LTP is indeed involved in declarative memory. To pin things down even better, mutant mice were manipulated genetically to have specific genes involved in LTP removed from their genomes. By "knocking out" the components of LTP, step by step, researchers were able to show that specific parts of the LTP process affect the declarative aspects of the spatial memory of the mice.

The second approach, using rodents with wires in their brains, can be used to study mice with proteins that are deficient in the LTP process. The experimental procedure is carried out by wiring a mouse's brain with a spike discriminator (a device that measures action potential in a specific cell of the brain). In this case, the probe is inserted into what is called a pyramidal cell of the hippocampus. These cells are particularly interesting because they are related to spatial recognition. The mouse awakes after the surgery with a wire coming out of its head. It is placed in a circular enclosure and allowed to explore, while spikes of action potentials are measured in the single cells where the probes are placed. The position of the mouse when spikes occur is recorded, and a map is made of where in the circular container the mouse is at the time of the spikes. It turns out that normal mice can locate themselves spatially, as indicated by individual cells that spike in only one small area of the circular enclosure. In contrast, mutant mice in which a gene involved in LTP is altered have hippocampal cells that fire randomly, with no correlation to spatial location within the circular container. This is yet another strong indication that LTP is involved in real living systems.

In the past couple of years scientists have started to pin down the biochemical and genetic bases of synaptic plasticity and LTP. Most of the studies we've just described show that, in long-term memory, consolidation of information happens very soon after the events to be stored are experienced. So much for initial storage; but how is that information maintained? The maintenance of LTP plausibly requires the presence of a long-acting protein, and one such protein has been discovered and named protein kinase Md (PKMd). Research on this protein and others is getting at the basis for other aspects of memory storage. Still, this isn't the whole story. Another approach to understanding the long-term potentiation needed for memory to work involves a model called the "synaptic tagging and capture hypothesis." This hypothesis attempts to explain how neurons commit to, and remain committed to, a particular potentiation. The idea is that synapses are somehow tagged and dealt with later in the process of memory storage. This hypothesis has accrued support from genetics and psychological experiments, and specific molecules that might be involved have been targeted. The tagging theory addresses one aspect of memory we haven't yet touched on—namely, that the system also allows for memories to be lost at a later time. In other words, tagging allows for the committal of a cell to a particular state to be postponed. This is necessary because, if every sensory input we ever experienced was retained and cells were permanently committed to these memories, then our neural systems would quickly run out of storage space. Some long-term memories can also be cleared at the later stage when memories get consolidated; the tagging hypothesis allows for another checkpoint earlier in the memory process, acting as a filter before a memory is passed on to be consolidated.

We mentioned another kind of long-term process called long-term depression—the flip side of LTP—as an important process in memory. Research in this area has focused on two major kinds of LTD, those involving NMDA receptors and those involving glutamate receptors. The LTD pathways are very complex but are being unraveled for both of these kinds of LTD. LTD has an extremely important role in memory, learning, and indeed pretty much any other cognitive response that requires flexibility.

When I Was Younger I Could Remember Anything, Whether It Happened or Not (Mark Twain)

How long-term memory is processed in our brains is still being worked out, but we can still say quite a lot about it. As we have seen, emotions are processed by several interconnected brain parts in that area of the brain we settled for calling the limbic system. Memories are also processed in these same areas. Neural circuits connect the hippocampus and amygdala, the major hubs involved in memory and emotion, to the prefrontal cortex. Using the brain imaging technique known as diffusion tensor imaging, or DTI, scientists have been able to visualize these connections in very fine detail. The DTIs show looping bundles of neurons connecting several areas of the so-called limbic system: the amygdala, hippocampus, cingulate cortex, and other areas. The pathways loop around the inner part of the brain where the limbic system resides, and many branches of these neural highways connect out to the prefrontal cortex. This connection to the prefrontal cortex is critical, as it allows the human brain either to dampen out the very basic emotional reactions or to come up with behavior that will respond to the emotion or memory. The connections to the hippocampus and amygdala are particularly important. The hippocampus is central to making memories, as the case of H.M. shows, and this area of the brain has many connections to the sensory areas that we discussed in chapter 4. Marcel Proust's madeleine cookie seems to be the favorite of many neurobiologists to demonstrate that odors are a big part of memories, but other senses clearly are, too.

Because any one memory may be the sum of inputs from many senses, the hippocampus is connected to many regions of the brain where these senses are processed. In addition, because your memories also often have emotional responses associated with them, your amygdala becomes an important hub for storage of your reactions to things that have happened to you in the past. The amygdala is therefore tightly linked into the system, and it more than likely filters memories based on our emotional responses to them.

One important final statement is that memory is very flexible. In the case of humans, it is not a foolproof system for retrieving a record of past events. Our memories of actual events may be not only hazy but also highly inaccurate. There is mounting evidence that our memories are influenced by such things as leading questions and prompting (something our criminal justice systems need to consider) and that while some aspects of our memories are quite good at extracting the general meaning of past events, they are not so good at the specific details. An example from personal experience will suffice to show how pliable memory really is. At a recent museum event, one of us was asked, "Where were you on 9/11/01?" He replied: "I was in my office at the AMNH that overlooks Central Park. I heard sirens and noticed jet fighters flying over the park. I remember it was a nice clear day out of my window." It turns out that these "recollections" were inaccurate on many fronts. For one thing, his office at that time didn't look out over Central Park. The actual office had windows, but they faced the interior of the museum. In fact, the building that he remembered being in on that fateful morning hadn't even been built in 2001. Why this quirk of our memories should be so, biologically speaking, is still a subject of great interest to scientists.

7 BRAIN EVODEVO

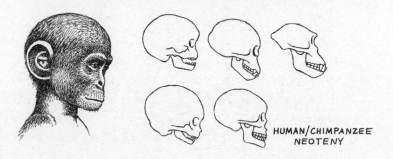

HUMAN/CHIMPANZEE
NEOTENY

Our brains and the brains of other organisms are continually changing, but how was the basic structure of the brain initially established? One way to approach this question is by examining the embryology of organisms, to understand the interplay of the genes and their environment in producing a fully developed brain.

Dirty Pictures

Remarkably, the embryos of all vertebrates show a high degree of resemblance. The nineteenth-century biologist Ernst Haeckel was the first to notice this, and in 1874 he published a now-classic diagram of how similar vertebrate embryos are, from fish to salamanders to turtles to birds to mammals and to humans (note the whiff here of the scala naturae; this should be a tip-off that something is amiss). Haeckel suggested that the embryonic development he saw proceeding from "lower" vertebrates such as fish to "higher" vertebrates such as humans reflected a process whereby evolution simply added more and more complex structures on to a basic plan. And he summarized his work in the snappy phrase "ontogeny recapitulates phylogeny." In one sense this statement meant that, when we observe the ontogeny (the development) of a series of vertebrates, we can reconstruct their degrees of relatedness using the development patterns observed for the most-developed embryo.

But Haeckel also thought that embryology was, more literally, a replaying of the evolutionary record. Creationists are quick to point out that Haeckel

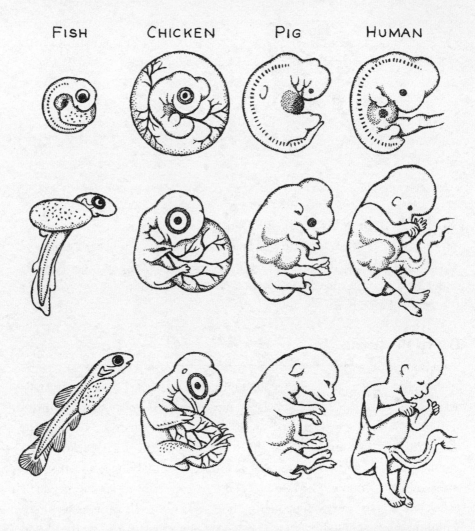

Ernst Haeckel's "ontogeny recapitulates phylogeny" figure, redrawn as in Joe Levine and Ken Miller's biology textbook. Three stages are shown in a fish, a bird, a nonprimate mammal, and a primate.

Brain EvoDevo

fudged his drawings to make the embryos look more similar to one another than they really were and thus feel they are discrediting the fact of evolution. However, in a widely used textbook, the biologists Joe Levine and Ken Miller point out that Haeckel didn't get away with the fraud in his lifetime. Several of his colleagues noticed discrepancies and demanded that he publicly admit he had fudged the drawings. Which he duly did—in a prime example of the scientific process at work! Nonetheless, when proper drawings of embryos are made, as Levine and Miller in fact did, the similarities still seem remarkable. We reproduce a version of this drawing here, so you can judge for yourself whether Haeckel's original idea (that a great deal of similarity is evident in the embryos of vertebrates) was correct; we think you might well agree.

Still, there is no need to suggest that evolution proceeds simply by adding changes to the end of already established developmental programs. As Levine and Miller expressed it, "Evolution can affect *all* phases of development, removing developmental steps as well as adding them, and therefore embryology is not a strict replay of ancestry." This is why an understanding of the underlying processes governing development of the brain is necessary. For it turns out that any stage of development can be altered and that even a loss of developmental steps can be responsible for present-day brain structures in organisms, especially vertebrates.

How the Brain Gets Its Stripes

How any animal develops from a single-celled egg-plus-sperm to a multicellular organism is also important when we are considering brain evolution. To get a better grasp of how this might happen, let's take a look at how the brain knows that it should partition itself into the three major divisions: forebrain, hindbrain, and midbrain. These are the "core" of the brain; add to them a spinal cord and a cortex, and you have the classic vertebrate brain. The key to understanding the process that gives rise to this complex structure is to understand that each cell in the developing embryo has the potential to become anything, because each cell carries the entire genome of the individual. The triggers that are in place to specify what a cell becomes in the developing embryo are the elements that control the differentiation of the different parts of the nervous system.

The genes involved in the early embryonic development of vertebrates including humans have been given strange names such as "Notch," "even-skipped," "engrailed," and "wingless." But why would a gene called wingless be important in the development of the human brain? Well, all the genes listed above were first discovered and characterized in the fruit fly, and the names are hence very fly-oriented. For instance, Notch refers to the phenotype of adult mutants for this gene—which have a notch in the wing—and wingless (wnt) refers to the phenotype that lacks wings. Engrailed and evenskipped refer to patterns of segment arrangement in early stage larvae of the fruit fly. And what is amazing is that all of these fly genes have homologues in vertebrate genomes. Further, they are all also what are called transcription factors.

Understanding the problem of the early development of the brain thus becomes one of discovering how these genes are expressed in the developing embryo. Tom Jessell, of Columbia University, points out that the early embryo can be viewed as a linear arrangement of tissues, rather like a tube. He also suggests that the developing neural tube is like a "tabula rasa where all of the cells are, at the start, equivalent." As we see in the illustration, the positions where genes are expressed along the tube can be visualized as different shades of gray on the tube. So discovering how the development of the brain occurs is like figuring out "how the brain got its stripes" with respect to these gray shades. And to characterize these stripes requires that we be able to see where and when genes are expressed, because gene expression determines the identity of these discrete brain region tissues.

Fortunately, scientists can view the expression of genes in embryos as they develop, thanks to a technique called antibody staining. The technique is simple, at least in its design. You make a molecule that will attach to a protein that is important in brain development, connect a fluorescent molecule to it, let it react with a fixed embryo, and look at the result under a microscope. Wherever fluorescence appears, the protein was produced. Using this approach, scientists have figured out that the differentiation of spinal cord, hindbrain, midbrain, and forebrain as distinct domains in the brain involves a gradient of protein expression of certain genes. In the case of the neural tube, signals are provided to the nascent nervous system from

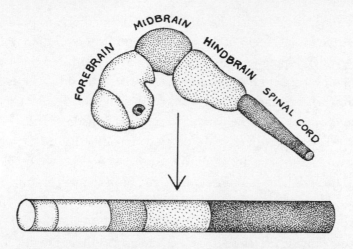

How the brain gets its stripes. The diagram shows the compartments of the developing nervous system in a prototypical vertebrate. These compartments are colored using shades of gray and translated to show how the developing embryo can be viewed as a linear arrangement of segments.

other cell types in the developing embryo. One kind of signal that is extremely important here is that of the wnt protein family. These are all signaling proteins with similar structures but with high degrees of specificity from one family member to another. They create a gradient that is at low concentration at the front end of the neural tube and at high concentration at its rear end. Because different arrays of genes are expressed in different cells, the gradient implements different cell fates. The target genes here are transcription factors themselves, and so a cascade of reactions is induced as a result of the initial gradient.

Wnt signaling is extremely important in the developing brain, with significant roles in neuronal connectivity. Thus wnt proteins control axon pathfinding, laying down the signals for the paths the neuronal extensions will traverse. In addition, they control certain aspects of synapse formation, axon changes, and the growth of dendrites in the nervous system.

Another important regulator of development in the developing brain and nervous system is called sonic hedgehog (as if the wacky fly names weren't enough). Sonic hedgehog (shh) also produces a gradient all along the neural tube, and in much the same way that wnt works to produce the basic tripartite arrangement of the brain, it controls the position of neural tissue in the developing brain. In fact, it has been estimated that over 40 percent of all early neural development is initiated by shh, so it is an incredibly important protein. In the head region, the shh gradient specifies where the eyes will sit in the nervous system. The gradient is highest farthest from where the eyes eventually develop, meaning that shh is produced in very low quantities or not at all in the area where the eye develops. If disrupted, this gradient can cause drastic anatomical changes. Imagine removing all shh gradually. Removal of a small part of the gradient—that is, making the concentration lower across the board—would result in the eyes shifting downward, closer to where the nose will eventually develop. Remove a little more, and the eyes shift even closer together. Remove shh completely, and the two eyes will overlap and actually be produced in the same place—making a Cyclops-like organism. Both chemicals ingested or applied from the environment or genetic mutations can cause this syndrome in some vertebrates.

Note that in both of these cases, gene expression is the key. How genes are turned on and off is a vital aspect of how the brain develops. The wnt and shh examples show clearly that sometimes the control involves a morphological gradient implemented by what is called a morphogen—in our examples, wnt and shh. Researchers have used this concept of gene expression control to explain some of the differences among our closest living relatives; with the sequencing of the Neanderthal (*Homo neanderthalensis*) genome, we are also beginning to see how gene regulation is different between us and this extinct relative.

The role of gene regulation in anatomical and behavioral differences between organisms was first noted by Allan Wilson and his colleagues at the University of California at Berkeley. Using a measure of protein similarity based on immunology, Wilson and Mary Claire King estimated that the divergence at the protein level between humans and common chimpanzees (*Pan troglodytes*) was about 1 percent. In other words, one out of every hundred amino acids in proteins differs between humans and chimps. Given the extreme anatomical and behavioral divergences of humans and chimps, this could only mean that the structures of proteins (that is, their shapes and sizes) had very little to do with the huge anatomical differences between them. So where did the anatomical disparities come from? King and Wilson hypothesized that how genes are regulated is the real arbiter of such larger differences among organisms. This incredible insight was made back in the 1970s; not until gene sequences from the genomes of humans and common chimps were compared in detail could the actual DNA sequence divergence be estimated. And it turned out that King and Wilson weren't too far off, because the human-chimp difference was measured at 1.23 percent at the DNA sequence level.

He Doth Like the Ape, That the Higher He Clymbes the More He Shows His Ars (Francis Bacon)

When King and Wilson undertook their original analysis, they examined almost fifty proteins, a Herculean task back in 1975. Ideally, to test their hypothesis exhaustively one would want to look at the expression of thousands of genes in different tissues. And doing this took the development of a technique that could examine the expression of large numbers of genes. This technique, called a microarray, is like having a little lab bench where you can conduct thousands of experiments simultaneously, and on a microscope slide to boot. The method allowed researchers to quantify which genes are expressed, and in which tissues they are expressed, in both humans and chimps. Several research groups have shown that, when overall gene expression is measured using microarrays, the major differences between humans and chimps can, as King and Wilson predicted, be traced to which genes are being expressed—most especially, in the brain. And in some cases the main difference turned out to be that the human brain overexpresses many more genes than the chimpanzee's does.

Indeed, in a review of the studies conducted up to 2004, Todd Preuss, Mario Cáceres, Michael Oldham, and Daniel Geschwind pointed out that human brain tissues overexpress from two to eight times more genes than the chimpanzee brain does. And another review of microarray data by Philipp Khaitovich, Wolfgang Enard, Michael Lachmann, and Svante Pääbo teased apart the dynamics of change even more. They compared gene expression in several tissues, including brain, kidney, heart, liver, and testes, and suggested that the brain showed the least amount of change among the tissues examined. In fact, the human brain has changed at most only half as much as the other tissues with respect to gene expression. What is significant, though, is that in this study the brain was the only tissue to show *accelerated* rates of gene expression change in the human lineage, thus agreeing with the conclusions of Preuss and his colleagues.

Even more detail about how gene expression affects the brain is provided by looking directly at the genes that are overexpressed and at how they are related to each other in their coexpression. When several protein products are examined in this way they form what are called modules, because they interact with one another. The protein connections in the modules can be characterized in humans and chimpanzees, and the striking result is that, in the brains of both, subcortical regions such the basal ganglia and the amygdala show stronger modular conservation than the cerebral cortex does. This result makes sense, because the major innovations in the evolution of our human brains are found in the cortical regions. Delving even deeper into differences in expression patterns between chimps and humans, researchers have homed in on two areas. One involves proteins that are active in energy metabolism, and the second involves transcription factors, leading to yet further refinement of King and Wilson's nearly forty-year-old hypothesis. The new information suggests that a small number of concerted changes in transcription factors (about ninety of them) have resulted in the major differences that we see in gene expression patterns between the brains of chimps and humans.

Another older idea about the evolution of chimp and human brains concerns the *timing* of developmental events. Stephen Jay Gould championed this idea in his book *Ontogeny and Phylogeny* (yes, the same ontogeny and phylogeny Haeckel proposed). Gould summarized the literature showing that timing events in developmental pathways could be critical in generating morphological, and perhaps even behavioral, diversity. As first developed by Haeckel, such changes are known as "heterochronic." Several outcomes of heterochrony discussed in the literature include one in particular that is of interest when discussing human-chimpanzee differences.

"Neoteny" is the mechanism by which adults of a descendant species retain juvenile features of an ancestral species. The outcome of neoteny is known as "paedomorphosis."

A classic natural example of paedomorphosis is the axolotl (*Ambystoma mexicanum*). As adults these curious salamanders retain the juvenile tail and gills and look like huge tadpoles. But their reproductive equipment is fully functional. *Homo sapiens* is sometimes times called the "paedomorphic ape" because some thirty traits in adult humans are equivalent to juvenile stages in the reconstructed ancestor of *Homo sapiens* and its ape relatives.

This overlap in traits occurs because the developmental trajectory of higher primates in general determines that juvenile stages are quite similar across species like gorillas, chimps, and humans. In the most obvious case, the chimp or gorilla infant skull starts out looking a lot like a human infant skull. During development in the apes, the face lengthens and slopes forward, whereas the adult human face remains flattish, somewhat like the infant stage. If you understand neoteny, then you are like Aldous Huxley, who used the principle in his book *After Many a Summer Dies the Swan*. In his book Huxley tells the story of a millionaire who seeks eternal life through the researches of a scientist. After a murder, some rare books, and an affair, Huxley arrives at a startling conclusion in which he describes the discovery of a European human being who has contrived to attain immortality. Unfortunately, this human has developed very apelike characteristics: apparently, the paedomorphic ape (us) develops into a regular ape if given enough time.

So much for literature! But brain gene expression studies have also weighed in on the classical evolutionary idea of neoteny in humans. Researchers examined the expression of genes in the cortex of humans, chimpanzees, and rhesus macaques at different developmental stages. In agreement with the notion that humans are, at least to some extent, neotenic apes, the researchers showed that the expression of genes in the cortex is delayed in humans relative to these two other primates. Furthermore, they show that the delay in expression is not uniform for all genes being expressed but instead affects a specific subset of genes that are important in the development and growth of neurons.

Return to the World of RNA

Examination of the control of gene expression in the brain has unveiled two important kinds of processes, relevant to brain development, that bring us back to the RNA world discussed in chapter 2. We described this world as one that was dominated by the single-stranded nucleic acid RNA. This versatile molecule can both carry information and catalyze chemical reactions; apparently, some of the remnants of the RNA world have been retained in our present-day DNA world.

The "central dogma" of the DNA world is that information flows one way: DNA to RNA to protein. Recently, though, it was found that the pool of transcribed RNA in cells contains small fragments of RNA that are just eighteen to thirty nucleotides in length. There are huge numbers of these in each cell, and their function is only now being deciphered. Collectively they are called small RNAs (smRNAs), and they are divided into several different types, based on their length and function.

The discovery of smRNAs was facilitated by the new "next generation" nucleic acid sequencing technologies, which can generate information on millions of individual RNA molecules during a single run. These approaches allow researchers to characterize the numbers and kinds of these RNA molecules that are floating around in cells; smRNAs have been found both to influence the regulation of genes and to be important in cell fate determination and stem cell biology. There are hundreds of specific kinds of these smRNAs in the human genome, each with its own function. So how do these little fragments of RNA regulate larger genes so precisely that specific tissues are highly dependent on them for proper development? Well, the role of smRNAs in regulating gene expression was first uncovered in the nematode *Caenorhabditis elegans*, and two examples will serve to illustrate the intricate and specific roles of smRNAs in neurogenesis.

The first involves "guilt by association." In this example, the processing of smRNAs, specifically the kind called miRNAs (microRNAs), has been examined in detail. miRNAs are processed using a series of proteins that involve the cleaving of the miRNA by a protein called Dicer. Once the Dicer protein has clipped the miRNA, the miRNA combines with messenger

RNAs that are the templates for proteins forming a miRNA-induced silencing complex (or RISC). The RISC is then further processed to cleave the mRNA for a protein such that its production is halted, thus silencing gene expression. In zebrafish, mutants for Dicer are associated with several developmental anomalies of the nervous system. For example, the neural tubes of affected individuals are defective and lack the midbrain-hindbrain boundary that is important in the proper development of the brain. What's more, by supplying a Dicer mutant embryo with a specific kind of miRNA, scientists can actually "rescue" the mutant. In the same way, the RISC complex can be disrupted by mutation in a protein called Argonaut that is responsible for the formation of the RISC. If the RISC is not formed in mutant mice, then the silencing of specific genes will not occur, and this gene expression anomaly will result in the failure of neural tube closure. The "guilt by association principle" therefore implicates miRNAs importantly and specifically in the development of the neural tube and the brain.

The second kind of example is provided by what we call "teeter-totter experiments" and concerns the control by miRNAs of transcription factors. We use the term "teeter-totter" because the idea of the experiments is to add a miRNA to a system and then to ask how the miRNA affects the up- and down-regulation of neuronal genes, the assumption being that the miRNA is affecting the activity of transcription factors. These factors are proteins that usually bind to DNA in the genome, and in so doing they regulate the expression or the transcription of the genes involved. The impact of specific miRNAs on the transcription of specific genes involved in neurogenesis can thus be assessed, and one specific miRNA, known as miR-124, has been examined in this way. This miRNA has been shown to be the most abundant smRNA in the central nervous system of both vertebrate and invertebrate adults. But just because it exists in large amounts in a specific tissue is not enough "guilt by association" to demonstrate its role in neurogenesis. So to assess the role of this miRNA in the development of the central nervous system, scientists developed a cell-culture test system in which they could get the cells to produce lots of miR-124. Next, they simply identified and counted the number of genes that were up-regulated (produced in greater quantities than in a cell not producing miR-124) and down-regulated (produced less than in a cell not producing miR-124). Using the microarray approach we discussed earlier, in which the activity of thousands of genes can be followed, researchers found that the miR-124 resulted in both the up-regulation of neuronal genes and the down-regulation of non-neuronal genes.

Another RNA-related process in the cell that more than likely affects neurogenesis is a process called RNA editing. This results in the alteration of cellular RNA by environmental influences after transcription has occurred. Such alteration of genetic material by environmental factors is commonly known as epigenesis. RNA editing has long been recognized as important in a lot of living systems, but it is one process that has increased in a stepwise fashion from vertebrates to mammals to primates, with *Homo sapiens* apparently having the greatest levels of it. Most RNA editing is catalyzed by gene products called ADARs (adenosine deaminases acting on RNAs). When tissues are examined for the presence or absence of ADARs, it turns out that they are preferentially expressed in the nervous system; more importantly, one form of ADAR is expressed exclusively in the brain. Guilty by association again!

As we've mentioned, *Homo sapiens* is the species with the greatest degree of RNA editing. Further, when RNA editing in humans is compared with editing in mice, over 90 percent of the difference turns out to be due to the processing of a single kind of RNA. This RNA comes from small stretches of DNA, repeated throughout the human genome, that are called Alu repeats. They are so named because they were first characterized using an enzyme, called AluI, that sliced specific sequences in the repeats, making it easy to identify them. Alu repeats are spread all over the human genome: there are about a million of them, found in both coding and noncoding regions of all twenty-three human chromosomes. Most of the Alu repeats in humans are found in the corresponding positions in the genomes of other primates, but curiously, there are seven thousand or so unique Alu elements in the human genome that do not appear in other primates.

This sets up an interesting guilt-by-association series: (1) RNA editing is most important in the brain, which leads to (2) humans show much more RNA editing than other mammals, leading in turn to (3) most of this increased editing is in Alu repeats, which are exclusively in primate genomes, and finally to (4) primates are the lineage that has experienced the highest order of cognitive evolutionary innovation. Connect the dots, and one might conclude that Alu repeats and RNA editing have been critical cogs in the evolution of cognition in primates. But they are not translated into proteins, and there are lots of them. So just how might they be involved in regulating brain development? The best guess is that they are somehow active in regulating genes that are important in neurogenesis. RNA editing occurs not only in Alu RNAs (which are noncoding—that is, they are not translated into protein) but also in coding regions (in genes that are translated into proteins). And it turns out that a large number of the coding regions that are RNA edited are neural-specific genes, which is hardly surprising because RNA editing is going on in a major way in the neural tissues.

RNA editing has also been suggested as a mechanism for how memories might be stored. In chapter 6 we described neural plasticity and how it relates to memory. In the classical way of looking at synapses, neural plasticity concerns synaptic strength. But remember that one of the problems of long-term potentiation (LTP) as a mechanism for how memories are stored was the time issue. Long-term potentiation allows for the maintenance and strengthening of a synapse, but the length of time it persists is problematic when memories are stored for very long periods of time. The potential role of RNA editing in memory storage is quite different and is based on the observation that transcripts from genes that encode enzymes for DNA repair are subject to RNA editing. One aspect of our genome that we do know is affected by the DNA repair genes is our immune system genes. Organisms with adaptive immune systems have suites of genes that control the production of the proteins active in immune response. The system senses environmental factors, such as viruses and bacteria and other causes of infection. Alterations in the immune gene sequences themselves are then made by the cell to deal with the infections in a process that generates what is called receptor diversity.

Receptor diversity is an important aspect of epigenetic genome change that permits the genome to cope with future infections. The whole process is called DNA recoding, and it results in a stable alteration of the DNA of certain cells at the immune system genes that make the antibodies which fight off infections. It is only a small jump to suggest that recoding might also occur in nerve cells; such recoding might in fact be the mechanism by which learned episodes and experiences are rewritten back into the DNA of the cell. What might happen is that a nerve cell would incur an epigenetic change in its genome (remember, all cells have a full genome's worth of DNA in them), which would tell the synapse to hold the potentiation and, hence, the memory. As the neurobiologists John Mattick and Mark Mehler suggested in a 2008 review, "Individual neural cells will in fact have distinctive spatially and temporally defined genomic sequences and chromatin structure," and "Memory as the consolidation storage and retrieval and long-term adaptations of human brain form and function should be modifiable by the targeted and differential modulation of expression of the genes encoding enzymes involved in RNA editing and DNA recoding." A very interesting idea indeed, but one that is based so far on guilt by association.

Size Matters (Or Does It?)

A hallmark of brains that develop higher forms of cognition is an increase in brain tissue real estate. In other words, the larger the brain, the greater its potential for higher cognitive functions. But it isn't all quite as simple as "bigger is better." Absolute brain size is important in various species of humans (see chapter 10), but brain size relative to that of other parts of the body is even more important when it comes to discussing increasingly complex cognitive function. One way to tease apart this relative size issue is through brute measurement of brain size in different organisms and then correlating this with some aspect of the biology of the organisms being measured.

Since the 1960s we have known that there is a rough correlation between brain size and body size in vertebrates. What this relationship means is that, as it develops, the brain keeps in scale with the rest of the body. But it's also frequently pointed out that there are some significant departures from this general relationship. Thus it has long been known that both dolphins and humans have bigger brains than would be predicted from their body sizes, based on the overall trend among mammals. In the past decade or so researchers have attempted to get at this observation in more detail and have achieved some interesting and unexpected results.

For proudly big-brained humans, it seems as if almost the biggest insult one can sling at a group of organisms is to say that they have small brains. In chapter 5 we mentioned the prevailing idea that birds have "feather brains." But in fact most birds have perfectly regular brain-to-body size ratios. And studies of bird species that deviate from this normal ratio illustrate how smart certain birds really are. Daniel Sol and his colleagues have looked at more than six hundred cases of the natural movement, or artificial introduction, of bird species into novel environments. They measured the brain-to-body ratios of all the bird species involved, both before and after the introduction events. And it turned out that birds with bigger brains are more successful at invading and establishing themselves in novel environments. In most cases, Sol and colleagues were also able to show that the success of these big-brained birds was due to their ability to innovate rather than through noncognitive mechanisms.

Another group of organisms that has an undeserved reputation for small brains is Marsupialia. This group of mammals lacks a placenta and is thus grouped, along with monotremes, among the nonplacental mammals. Vera Weisbecker and Anjali Goswami have revisited the brain sizes of marsupials and placental mammals, but with a twist. Most comparisons ignore the basal metabolic rate, a measure of the rate at which the body takes in and processes nutrients. But Weisbecker and Goswami factored this in. They also examined the time that the mothers devote to taking care of their young from birth. When these parameters are included, it turns out that basal metabolic rate and maternal investment are actually more important than body size in influencing brain size. So, although marsupials have small brains, it is not for the same reasons as among small placental mammals.

This idea that social characteristics such as maternal care can influence body size has also been examined in several placental mammal groups. Because higher sociality requires more communication skills and greater cognitive abilities than solitary lifestyles do, social behavior has been thought to require more brainpower. This is sometimes called the "social brain hypothesis." Our American Museum colleagues John Finarelli and John Flynn examined the phenomenon of increased brain size and sociality in one of the sexier orders of mammals, the carnivores (Carnivora: dogs, cats, bears, and so forth). Using only living species, they detected a correlation between sociality and brain size. But when they used both living and extinct species (four hundred, of which about one-third were fossils), they found that changes in brain size were plentiful, occurring many times independently in carnivore evolution in groups with different social habits. So increase in brain size is more than likely much more complex than just a function of increasing sociality. In fact, when the social brain hypothesis was examined directly using fossil data, there was no evidence at all for correlation of socialization with larger brains. This is in line with other studies using more precise parameters than simple socialization. One such study, by Simon Reader and colleagues, used what they called "ecologically relevant" measures of cognitive ability in an attempt to see if they correlated with increased brain size. These researchers examined about five hundred cases each of innovative behavior, social learning, and tool use among animals. And when they checked these ecologically relevant indicators of cognition against brain size, they found a correlation only when they used very precise definitions of social characteristics.

Microcephaly

Although some of the studies we have considered so far shed doubt on the direct correlation of brain size with cognitive abilities, one group in which the trend is pretty clear is in the group to which we humans belong: the primates. The more derived a primate lineage is, the larger seems to be its brain—and in particular, the neocortex. What's more, multiple lineages of primates have experienced independent increases in brain size. And finding multiple independent events in evolutionary history is an evolutionary biologist's dream, because comparison of a similar phenomenon across several lineages allows study of the particular evolutionary event in detail. Together with the increasing ease of analysis of primate genomes, this gives us the opportunity to use multiple lineages as a tool for understanding genetic and genomic effects on brain size enlargement.

One group of genes involved in brain size increase is curiously named "microcephaly" genes—curious because the name implies "small brains," but less curious, perhaps, when one considers that these genes were found as a result of studying a collection of human disorders lumped under the term "microcephaly." This syndrome is a congenital disorder in which the cerebral cortex does not grow to its proper extent, and there are apparently no environmental, chromosomal, or metabolic anomalies involved in it. Geneticists have determined both that at least eight genes are involved in the expression of the syndrome and that it is inherited. Four of the genes examined so far make proteins involved in the formation of complexes affecting cell division in nerve cells. These genes affect processes that are important in cell division, when the newly replicated and old sets of chromosomes move from the center of the dividing cells into their new daughter cells. These four genes are thus most likely involved in regulating the number and kinds of cells that are available for further division. Another gene is involved in the DNA damage-response pathway of the cell. When cells divide, they automatically check their chromosomes for damage and then attempt to correct any they find before replicating. If damage is not detected or isn't repaired, then the dividing cells are in jeopardy of not being able to divide. The final three genes of the eight that are implicated in microcephaly have not yet been characterized.

How brain tissues expand during development necessarily relies heavily on cell division during embryogenesis. Most of the hypotheses that attempt to explain this tissue growth, and hence how the shape of the brain is "sculpted," rely on differential symmetric and asymmetric cell division in the brain, most likely implemented by exactly how cells divide during mitosis. Via such mechanisms, the genes we've just been discussing are plausibly involved in how brains get bigger. Certainly, the unfortunate babies that have mutations in these genes are born with considerably smaller than normal brains. These infants may live for some time, but they are mentally challenged. When the brains of these microcephalic individuals are examined, an interesting observation emerges. One might think that decreased size might result in reduced thickness of the cortex. After all, in microcephalic individuals the cranium is smaller and more compact. But it turns out that thickness is not reduced. Instead, the surface area of the cortex is reduced. And if we think about the roles of the genes involved this makes a lot of sense, because almost all of them regulate the size of the pool of nerve cells available for division—and this is entirely consistent with the reduction of the surface area of the cortex.

Looking at the brain from the right perspective is hugely important when we are considering how brain size has changed in the history of organisms. And knowing that these five microcephaly genes are somehow involved gives researchers a better means to evaluate brain size in primates. So although most studies to date have looked at overall brain size (including cortical thickness, by default), it might actually be better to look at cortical surface area, or at the brain mass, across a broad range of primates. Brain mass and cortical surface area are in fact correlated much better with the total number of neurons than they are with absolute brain size, and it is exactly this number of neural cells that is affected by the genes we have been discussing.

Several research groups have suggested that the four genes that affect mitosis directly, by altering structures that implement chromosome movement during cell division, are under intense adaptive selection. And they believe that the selection has been especially forceful in the lineage leading to us. In other words, they believe that our bigger brains are the result of natural selection for changes in these genes. But an examination of this phenomenon in primates by Stephen Montgomery and his colleagues suggests that only two of the genes may be involved and that similar selection has occurred at many places in the primate phylogeny. More important, Montgomery and colleagues conclude that cortex size is not the limiting factor in primate brain size evolution; rather, the surface area and mass of the brain are impacted in the lineage-specific changes of brain size in primates.

Changes at the genome level to implement alteration of brain anatomy, of the kind we see in microcephaly, are fundamentally different from the kinds of genetic changes that seem to have occurred most widely in the evolution of the brain. Remember the work King and Wilson did, and the subsequent research with microarrays that showed gene regulation to be predominantly responsible for changing brain anatomy and physiology in the chimpanzee and human lineages. The genes we have just discussed are in fact among the relatively few good examples we have of evolutionary change implemented by alteration of structural genes—the genes fighting in the trenches, as it were, in contrast to the more ubiquitous changes implemented by their commanders, the regulatory genes.

8 WORDS AND MUSIC BY . . .

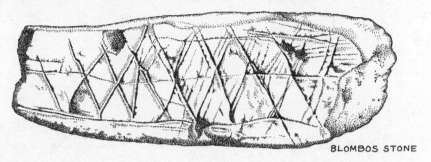

BLOMBOS STONE

There are certain things the human brain does that are clearly unique. Among these are language and music. How do such important aspects of our so-called higher cognitive abilities work, and how did they evolve?

Language Areas and the Area of Language

Language is more than just communicating. If it were just that, then clearly a lot of species, if not every species on this planet, would have it. René Descartes actually recognized this conundrum when he suggested (rather judgmentally) that animals could not be conscious rational beings because they do not have language. But the fact is that numerous animals have the ability to communicate vocally in sophisticated ways. Dolphins and vervet monkeys, for instance, communicate using vocal mechanisms that are extremely efficient and even elegant. And there is a bonobo called Kanzi who can deploy at least some aspects of language for communication. Such realities make "language" a slippery word: like many of the words we have examined in this book (such as "brain," "synapse," or "homology"), if we twist it around enough "language" can mean almost anything. It is thus important that we define what we mean by language right away, before we look at its origins in the brain. Here are some descriptions we encountered in our efforts to discover a definition:

> The words, their pronunciation, and the methods of combining them, used and understood by a community
>
> A particular kind of system for encoding and decoding information
>
> A formal language is a set of words, i.e., finite strings of letters, symbols, or tokens
>
> A form of communication using words either spoken or gestured with the hands and structured with grammar

The emphasis varies, focusing either on words or on the rules for combining them. But the key, we think, is that both aspects are involved: all of the world's six thousand languages have in common that a finite vocabulary of (usually) auditory symbols can be combined, according to rules, to produce an infinite number of statements about the world. We spend a lot of our lives talking to each other, so we intuitively think of language as a uniquely precise and flexible means of communication—although, of course, with about six thousand languages in the world today, each of which has to be specifically learned, language is actually one of the greatest barriers there is to communication. Indeed in this respect it is all but emblematic of the cultural differences that both beset us and make the world such an interesting place. But one of the more important aspects of language is that it permits us to think in the unusual way we do. Language provides us with a way of categorizing the world around us into discrete abstract symbols that we can shuffle in our minds, and it gives us the organizational rules that we need to link and rearrange those symbols into coherent ideas. Without language we would be unable not only to communicate as we do, but also to think so uniquely.

There are four basic approaches to studying the evolution of language. The first uses the tried-and-true methods of cognitive neuroanatomy that resulted in the localization of speech functions to Broca's and Wernicke's areas. Delimiting such areas can also help us interpret brain casts of extinct hominids, and the modern descendants of this line of attack include amazing imaging techniques that weren't available to Broca or Wernicke, or even to Wilder Penfield in the mid-twentieth century. These methods include functional magnetic resonance imaging, positron emission tomography, and magnetoencephalography.

The second major approach is to use genomics, much as those microcephaly genes were used to understand the underpinnings of brain size (chapter 7). The third is to study the relationships of languages themselves, using specific language components as traits for reconstructing the history of known languages. And the final approach is to fall back on other systems to help us decipher the evolution of language. But whichever approach we adopt, we always have to keep in mind exactly what it is that we mean by language, to keep the term from falling into the metaphor bin.

Regions of the brain involved in language have been extensively mapped using the clinico-anatomical method adopted by Charcot almost two centuries ago. And it should not be surprising that something as complex as language turns out NOT to have a single location in the brain. Nonetheless, some of the brain regions involved in producing language are themselves quite localized: you use one area when you express yourself and another when you're interpreting what someone else is saying. And of course, all of the various brain areas involved work together as a single integrated network. One of the more interesting results obtained from attempts to localize language to specific areas of the brain is the right-left dichotomy that we have already discussed in the context of vision and body control. At its grossest level, language can be localized to both sides of the brain. But although the left side seems to be responsible for handling the literal meaning of words, the right side is engaged with more subtle things, like understanding emotions that might be embedded in language (sarcasm and humor, for example, or the detection of tone of voice). What is perhaps most interesting is that understanding metaphor is also a function of the right side of the brain.

Some spectacular conclusions about the left-right brain dichotomy have been made from people who have had communication between the left and right sides of their brains physically cut off, in what is called the "split brain phenomenon." These people are usually epileptics who have had their corpus callosum severed surgically to relieve seizures in a procedure called commissurotomy. The corpus callosum is the midline brain region, literally a bundle of nerves through which all communication between left and right sides occurs. People who have had commissurotomies manifest strange perceptions of the world, and these have allowed neuroscientists to tease apart some of the intricacies of the control of language. In split-brain experiments conducted first by Robert Sperry and more recently by Michael Gazzaniga, images of objects were projected consecutively to the subject's left and right eyes, by means of a device with a divider that ensured the images were perceived by only one eye. At the same time, four cards were presented to the subject's left hand and four to his or her right. These cards did not necessarily directly coincide with the projected pictures, but they were related to the projected scenes in some way.

In one of Gazzaniga's experiments, he presented a snow scene to the left eye and a chicken foot to the right. Because of the way the visual system is wired, the left eye sends information to the right hemisphere and the right eye to the left hemisphere. When any subject picks the corresponding picture for a specific test, his or her right brain will respond with the left hand, and the left brain will respond with the right hand. The split-brain subject will easily identify a card related to the chicken claw as such with the right hand and will correctly identify a card associated with the snow scene with the left hand. But by manipulating the pictures that their subjects used to identify the snow scene and the chicken claw, the researchers cleverly revealed interesting aspects of the split brain. Thus, if the snow scene was identified with a shovel (for clearing the snow), the subject would refer to the shovel as the picture best associated with the snow scene. If the subject was then asked why his or her left hand pointed to the shovel, there would be no response from the right hemisphere. The left hemisphere, however, also understood the question, and since this is the side of the brain that dominates for language and it had seen the chicken, the subject would make up a story incorporating what it saw: an explanation for the chicken. One subject accordingly pointed to the shovel because it could be used to clean up a chicken coop. Results like this point to the importance of communication between the right and left sides of the brain.

Why did the subject in the last case make up the logical but silly story about the shovel and the coop? The researchers' reasoning about what was going on went like this. The initial decision to point to the shovel was made by the right side of the brain, which saw the snow scene. But the brain's right side is not good at speech or problem-solving and remained stumped and speechless—hence its lack of response on seeing the snow scene. Since the left side of the brain dominates for speech, the subject answered with the best response he or she could, based on what that side of the brain saw: a chicken claw.

Through many years of work, starting with Broca and Wernicke, researchers have arrived at a pretty good picture of the language areas of the brain and how they function in producing, understanding, and interpreting language. The American Museum of Natural History's *Brain* exhibition summarizes these major functional areas as follows:

> PUTTING WORDS TOGETHER—*Broca's area:* Generating the words you want to say or write uses this region. People with damage to this area can understand what others are saying but have a lot of trouble answering.
>
> TALKING AND WRITING—*Motor cortex:* Specific areas in your *motor cortex* control the muscles in your mouth, tongue, and larynx when you speak and the muscles in your hands when you write.
>
> UNDERSTANDING LANGUAGE—*Wernicke's area:* Understanding the speech and writing of others takes place here. People with damage to this region cannot understand what others are saying. They can still talk, but their speech may be jumbled.
>
> HEARING—*Auditory cortex:* The sounds of others talking are first processed in the *auditory cortex.* The message is then sent to Wernicke's area, which figures out what the sounds mean.
>
> SEEING—*Visual cortex:* Written words are first processed in the *visual cortex.* But the meaning of the words and sentences is decoded later, in Wernicke's area.

This doesn't sound too complicated. But when you add the intricacies of other aspects of language and where they are localized in the brain, the picture begins to get convoluted, and you realize just how complex language is. Consider this: in the first nine months of 2009 alone, a hundred publications using imaging approaches to localize aspects of language to specific brain regions appeared in the scientific literature. The neurobiologist Cathy Price summarized these studies in a 2010 review, correlating about twenty well-defined areas of the brain with various aspects of language, including: prelexical speech perception, detection of meaningful speech, semantic retrieval, sentence comprehension, recognition of incomprehensible sentences, prior knowledge of semantic associations, word sequences and word articulation that predict the content of a sentence, speech production, word retrieval, articulatory planning, execution of speech, and suppression of unintended responses. She concluded that there are indeed specific brain areas playing major roles in many of these complex language tasks.

Given their complexity, there is inevitably some controversy as to just how useful these imaging approaches are at localizing specific functions in

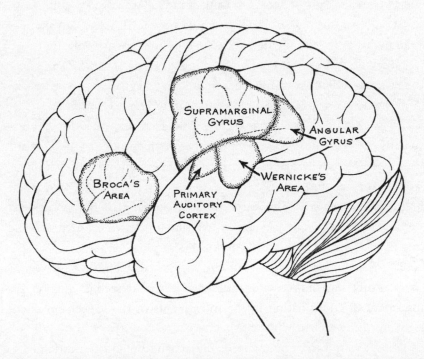

Language regions of the brain, showing the major areas involved in language.

language. The major criticism is that the imaging data are collected for groups of people and that variability in the linguistic capabilities of the individual people scanned obscures the localization of the language functions. In an interesting critique of this approach to the localization of the elements of language, Evelina Fedorenko and Nancy Kanwisher asserted that "two broad questions have driven dozens of studies on the neural basis of language published in the last several decades: (i) Are distinct cortical regions engaged in different aspects of language? (ii) Are regions engaged in language processing specific to the domain of language? Neuroimaging has not yet provided clear answers to either question." But another neurobiologist, Yosef Grodzinsky, took issue with this assessment: "When the neurolinguistic landscape is examined with the right linguistic spectacles, the emerging picture—while intriguingly complex—is not murky, but rather, stable and clear, parsing the linguistic brain into functionally and anatomically coherent pieces." Clearly, the last world on this is far from being written. But one thing is for sure: however revealing the functional magnetic resonance, positron emission tomography, and magnetoencephalography localization approaches may prove to be, the task of mapping the various skills involved in language to the various parts of our brains is turning out to be stunning in its complexity.

FOXP2

Many genes are expressed in the brain, both during its development and during its maintenance in life. For one, the microcephaly-related genes already mentioned give researchers a nice tool to approach the understanding of brain size in animals. And fortunately for language studies, there are several gene candidates in this area, too. One of the more promising of them is called FOXP2. This gene's involvement in language and speech was identified through the study of a family known as KE. Half of the members of this family showed a syndrome called verbal dyspraxia, which is an impairment of coordination in the movements of the speech-production system resulting in individuals who have a very difficult time speaking. After extensive testing of speech parameters and behaviors in members of the KE family, researchers concluded that the disorder had the greatest impact in the areas of word-nonword repetition and orofacial praxis (volitional control of skilled nonspeech facial movements). Many other speech-related traits are also affected, but these two unambiguously allowed researchers to examine this disorder. As Faraneh Vargha-Khadem, David Gadian, Andrew Copp, and Mortimer Mishkin pointed out, "One core deficit in affected family members is a higher order orofacial motor impairment or dyspraxia that is best exemplified in speech, because speech requires the precise selection, coordination and timing of sequences of rapid orofacial movements." This syndrome presents aspects of speech behavior very similar to the Broca's aphasia we've already mentioned, although whereas the KE family syndrome is specific for both word-nonword repetition skill and orofacial praxis, Broca's aphasia affects only word-nonword repetition. Note at this point that both Broca's aphasia and the KE family syndrome affect very localized aspects of speech and language production, so that we are talking about a small part of the overall phenomenon of language. And although identifying the single agents of disorders is an excellent way to begin understanding the whole of a behavior, even if we are successful at deciphering the single gene for the localized defect, we haven't discovered the gene for the whole process.

Genetic linkage analysis was performed on the KE family, and the syndrome was localized to chromosome 7, a rather run-of-the-mill chromosome of medium length. And at this point an unrelated individual, CS, who had the same signs of the disorder, was brought into the analysis. Luckily, CS had a chromosomal translocation about which the doctors already knew and suspected of involvement in CS's speech problem. This translocation had a breakpoint (the point on the chromosomes where the translocation occurred) that could be localized easily using molecular cloning techniques. Researchers then found that the breakpoint was in a previously unnamed gene. When they looked closely at this gene using DNA sequencing, they discovered that it was very similar to a *Drosophila* gene called "forkhead" that affects development of the fruit fly gut.

Forkhead proteins have a specific amino acid sequence embedded in them that makes them important in gene transcription and regulation during development. As it happens, in mammals there are many proteins with forkhead domains in them. The specific one that was discovered to contain the breakpoint in CS's genome is called forkhead box P2 (FOXP2). The FOXP2 gene codes for a protein that is more than seven hundred amino acids long. This protein binds to chromosomes in which it regulates genes. When the KE family was reexamined for the gene sequences of the FOXP2 gene, an odd change was seen. Position 553 in normal individuals is filled by the amino acid arginine. In affected members (only) of the KE family, this arginine is replaced by a histidine. Position 553 in FOXP2 is vital to the protein's function in regulating other genes, and it lies near the critical part of the protein that binds to the chromosomes it regulates. A lack of arginine in this position means reduced or improper binding.

Researchers then examined the neuroanatomy and function of the brains of KE family members, using functional magnetic resonance imaging and an analytical approach called vox-based morphometry. These studies localized differences between affected and unaffected individuals of the KE family to a specific part of the basal ganglia known as the caudate nucleus. This area of the brain is a cluster of nerves that is important not only in memory but also, in conjunction with the thalamus, in language comprehension. In addition, the structural studies uncovered unusually low amounts of gray matter in Broca's area (and several other localized parts of the brain), while levels of gray matter were higher than usual in Wernicke's area and a structure called the putamen. Functional magnetic resonance imaging studies of the brains of the KE family members affected by the disorder showed lower-than-usual activity in Broca's area and the putamen—but no such pattern in the caudate nucleus. Of note is that the affected individuals concomitantly showed overactivation of regions of the brain that aren't usually associated with language and which the researchers thought might reflect the added effort the affected individuals needed to apply when completing the tasks they were given during the resonance imaging tests.

Putting all this information together led researchers to suggest that FOXP2 is involved in wiring the circuitry in specific areas of the brain. In fact, when they examined where in the developing fetus FOXP2 is expressed, they found significant overlap of expressed regions with the anatomical areas just discussed. An electrician wiring a house for lighting knows the parts of the house that need to be fitted with light fixtures and wall switches, and then he or she finds the best conduits to allow electricity to flow efficiently to the specific locations of the lights and switches. Likewise, Vargha-Khadem and her colleagues tried the same approach with the wiring of the brain in the FOXP2 system. They suggested that: "by integrating the evidence about the neural phenotype of the KE family with the neural expression pattern of FOXP2 the proposed circuitry provides a tentative account not only of how the KE mutation has resulted in the affected members' orofacial and verbal dyspraxia, but also of a way in which the normal FOXP2 gene might have contributed to the emergence of proficient spoken language." And this, of course, brings us back to evolution.

Various vertebrates have been surveyed for this gene, and the results are as one might expect in the light of knowing that it is already important in *Drosophila* development (though there it's in the gut). We find FOXP2 in distant relatives such as birds, in close relatives like chimpanzees and gorillas, and even in Neanderthals. What is surprising, though, is that in birds the protein is almost identical to the human protein, differing at only eight positions between the two. And the strange identity extends into other realms, because the places where the gene is expressed in the brains of mammals and songbirds are also remarkably similar. We mentioned earlier that FOXP2 is not the only protein with the forkhead domain; indeed, a protein very similar to FOXP2 is called FOXP1. In birds, the expression of FOXP1, which binds to FOXP2, is much greater in males who are learning songs than in females who don't learn songs.

Bats have an equally interesting FOXP2 story. Some bats can echolocate using sonarlike mechanisms in a rapid, fine-tuned process. It turns out that echolocating bats have changes in the same region of the FOXP2 protein that *Homo sapiens* has and that nonecholocating bats don't! These nice results lead to two conclusions. First, they strongly suggest a correlation of FOXP2 and FOXP1 with the production of songbird song. Second, they make it seem very likely that, at some level, FOXP2 has an important role in human speech production. The bat results are significant in respect to this last point, because the processes of echolocation and speech both require finely tuned motor skills; perhaps the FOXP2 gene controls the tweaking of these skills to produce echolocation in bats on the one hand and speech in humans on the other.

In both chimpanzees and gorillas the number of sequence changes relative to humans is on average two, occurring at positions (304 and 326) in which only *Homo sapiens* shows deviations from the primitive condition. This might seem to make us very similar indeed to our closest living relatives—but in turn they differ only in one position from very remotely related mice! Even more remarkably, Svante Pääbo and his colleagues at the Max Planck Institute in Leipzig have recently deciphered the genome of *Homo neanderthalensis*, revealing that Neanderthals are identical, to the very residue, with humans.

Here is the logic that is usually applied to this amazing observation. FOXP2 has a role in speech. Humans differ from nonhominid primates at two positions in the FOXP2 protein. So, since Neanderthal FOXP2 proteins are identical to *Homo sapiens* FOXP2 proteins, Neanderthals as well as *Homo sapiens* probably had speech—and, by the researchers' calculations, had had it for about half a million years. Underlying all of this is, of course, a timeframe. Neanderthals and *Homo sapiens* probably diverged well over half a million years ago (see chapter 10). This would mean that the FOXP2 changes found in the *Homo sapiens* lineage today arose before the divergence of *Homo sapiens* and *Homo neanderthalensis*. Fair enough. But the suggestion that both had language really does not coincide with what we may reasonably infer about the Neanderthals from other evidence. So, given that the Neanderthal genome seems to be telling us pretty plainly that the FOXP2 protein was present in the common ancestor of *Homo sapiens* and *Homo neanderthalensis*, we might more properly conclude that, although FOXP2 may be necessary for language, it is not sufficient. Neanderthals may have had a very demanding style of vocal communication that required the FOXP2 mutation, but it was nonetheless not what we might recognize as language.

FOXP2 is a very stable gene among the vertebrates and is clearly highly resistant to change. So why did the human FOXP2 mutation ever become incorporated into the *Homo sapiens/neanderthalensis* lineage? After all, no other primates on the face of the earth have these two positions altered, and the change might potentially have involved some strain on the fitness of the ancestor that acquired the changes. However, it doesn't appear that this was the case. Remember that populations at this time in the history of the genus Homo were incredibly small. And small population sizes enhance the probabilities that the bizarre and less useful will become fixed. So if the FOXP2 alteration in the common ancestor of Homo sapiens and Homo neanderthalensis had been in the least bit advantageous for a function other than speech that we haven't yet figured out, it might have lingered on in populations of Homo neanderthalensis and Homo sapiens for reasons unrelated to speech. Or maybe it just hung around because it wasn't disadvantageous. In any event, although it is pretty clear that the human variant of FOXP2 is an important underwriter of our speech capacity, its presence in Neanderthals suggests that it was probably not responsible for the final stage of fine-tuning of our peripheral vocal systems.

A final possibility that has been mooted about language and modern behavior patterns is that, some fifty thousand years ago or thereabouts, another gene or set of genes changed that allowed the great cognitive leap forward on which we are so wont to congratulate ourselves. Making propositions of this kind can be useful if doing so involves really specific and testable hypotheses—certainly, if the proposers are ready to discard them if necessary. But as we pointed out in chapter 1, it is usually problematic to equate a gene with a behavior—though this hasn't stopped a lot of researchers from doing so.

Casting the Net for a Tree of Languages

One certainty is that, once our species had found language, the ability for efficient communication it conferred became the catalyst for many other major cognitive advances. And equally clearly, that acquisition was supremely recent on the evolutionary timescale. But to judge by the huge variety of languages and language families existing today, once language had taken hold, its diversification was extensive and rapid, a process that was probably hugely amplified by the isolation and tiny sizes of the Homo sapiens populations that existed during the first 95 percent of our species' existence. Ethnologue, a publication of the Summer Institute of Linguistics, records nearly 7,000 languages currently spoken on the planet. And this large number of languages pales in comparison with the number thought already to have gone extinct. Under current conditions, of course, that extinction is accelerating. It is estimated that, by the year 2050, 90 percent of the 7,000 or so languages that now exist will have disappeared. Even now, 500 languages are estimated to be spoken by ten or fewer people and are clearly on the brink of extinction. The situation is so dire that the Living Tongues Institute for Endangered Languages has cited five areas of the world as "hotspots" of language loss. These include northern Australia, northwestern and southwestern North America, the Amazon basin, and the far northeastern part of Asia.

The huge number of languages spoken in the world is a tribute to the importance of language in the success of *Homo sapiens*. And by looking at the relationships among languages, we can learn a lot about the history of how this important innovation affected, and continues to affect, the success of our species. One problem with reconstructing the histories of organisms— and in this case of languages—is that if the phenomenon you are following evolves too fast, then the probability of getting convergences increases enormously. One of the strategies used to avoid those convergences is to cherry-pick the information; but this is clearly not an optimal way to proceed. An alternative is to build a model that will allow one to "filter out" convergent and hence confusing information.

To facilitate the study of the world's languages, *The World Atlas of Language Structures* (*WALS*) was created. This atlas contains information on nearly 3,000 languages, and on 141 language features. Two major kinds of information are usually brought to bear when reconstructing the history of languages. The first category of traits is called "lexical," and it involves the comparison of words and their definitions across languages. The second category of traits is called "typological," and these include structural features such as sentence complexity, phonology, and nominal syntax. The language expert S. J. Greenhill and several colleagues took such data and constructed a phylogenetic tree that is much like the biological trees we discussed in chapter 1. Their hope was to discover categories of language characteristics that might be changing very, very slowly, because if they could find characters like this, then they might be able to reconstruct the early history of languages that has been lost in the cacophony of change. These authors' analysis is one of the more complete attempts to determine the relationships of languages and, more important, to estimate when the "mother of all languages" existed.

The tree generated by Greenhill and colleagues makes a lot of sense with respect to many previous studies, uniting languages that we have long suspected are related, such as the Indo-European group of languages that includes English, German, French, Russian, Latvian, Greek, and Spanish. And their outstanding finding, also echoing other studies, is that a tree is actually not the best way of describing the relationships among languages. Some groups of languages indeed cluster together nicely, but at the base of the "tree" there has been so much exchange of linguistic bits and pieces that it is better described as a "net." As these authors suggest, "Our results show a substantial non-tree-like signal in the typological data and a poor fit with known language relationships within the Austronesian and IndoEuropean language families."

As noted, Greenhill and colleagues tried to tease apart those language characteristics that have changed rapidly and those that have changed slowly in different parts of the world. And their findings in this respect are also interesting. First, they found no difference in the rates of lexical versus typological characteristics, although the lexical characteristics were much more variable. But when different categories of typological characteristics were examined, some very distinctive differences emerged in various parts of the world. For instance, sentence complexity, a typological characteristic, showed no difference in rate of change when Austronesian and Indo-European languages were compared. But a highly significant faster rate of change in nominal syntax, yet another set of typological characteristics, was found in Austronesian compared to Indo-European languages. These results indicate that there is no single typological linguistic category that evolves more slowly than the lexical category. So in the end Greenhill and colleagues concluded that their work simply "highlights how little we know about the shape and tempo of language change."

Although these results leave us with a problem that will not be solved without considerably more modeling of how languages change, the more we learn about the behavior of language characteristics, the better we will understand the models we might use. This is important, because understanding where and when the languages of our species originated is an important step in understanding our humanness—so it might be worth the wait to get this right.

Everybody's Talking at Me (Harry Nilsson)

Organisms have evolved myriad ways of communicating with each other. We have already addressed at length how single-celled organisms communicate at the cellular and molecular levels. But animals communicate in an altogether more complex manner, and they have evolved a huge variety of ways of doing so, using olfactory, tactile, and visual as well as auditory approaches. Visual means of communication rely on coloration or on movements that are stereotypical of a species. Coloration changes are well known among some fish, such as the African cichlids. Several hundred species of this group have formed in three of the largest of the East African lakes (Victoria, Tanganyika, and Malawi) over a very short period of time (from 130,000 years in the youngest lake to 2 million years in the oldest lake of the chain). With this many species around, it is important that each individual be able to communicate with its conspecifics both for purposes of mating and to ward off unwanted suitors from other species. The visual capacities of these fish have also evolved to tell the fish what is food and what is fiend (that is, what will attempt to eat them).

Fish, of course, live in water, and although sounds can be used to communicate in this medium, sight seems to have been the system of choice among the cichlids. The East African cichlids have an amazing array of coloration patterns for communicating species membership and for conveying predator-prey information to the viewer. The colors that these fish can see run the gamut of the wavelengths of visible and ultraviolet light. And it appears that, to increase their "vocabulary" (used here as a metaphor), these fish have evolved a complex array of opsin genes, the genes that make proteins that allow the rods and cones in the eyes to detect color.

Some animal species, among them fish, reptiles, crustaceans, and cephalopods, can change color and light polarity using organs embedded in their skin called chromophores. A wide array of chromophores can be used to alter coloration, so that an even bigger visual "vocabulary" can be employed by these forms. In most organisms color changes are controlled by hormones, which are sent to the skin to trigger the color change. In cephalopods, structures called chromophores are found in the outer layer of the skin, but other organs (called iridophores and leucophores) that lie in the skin below the outer chromophore layer also allow some animals (especially cephalopods) to reflect light or to produce polarized light. And, very important, the chromophore system is connected to the individual's nervous system, allowing for more rapid and precise control of the process of color change.

Cephalopods can use their ability to change color to protect themselves from predators by blending into their background. And if that weren't enough, some cephalopods such as cuttlefish can actually change the texture of their skin, adding even more to a potential vocabulary. All this is remarkable, and it makes great television. But researchers have also suggested that cephalopods use coloration change to communicate with each other beyond camouflage and the usual sexual banter of species. In an examination of the possibility that coloration changes might constitute more specific communication, Roger Hanlon and colleagues at Woods Hole Oceanographic Institute have pointed out that not only do cephalopods produce polarized light but they also respond to it. Hanlon is careful not to jump to conclusions, and although he and his colleagues are intrigued by the possibility that these color changes and light polarizations constitute communication, they do point out that the evidence for this is circumstantial. As a result, they have come up with a useful set of criteria to be fulfilled when going beyond circumstantial inference. For communication to be recognized, several things have to be observed: "(1) the production of a signal by a sender, (2) sensing of this signal by a receiver and (3) a change in the receiver's behavior and it must be beneficial to both the sender and the receiver." These are pretty good rules to follow when you are trying to decide if a species is really communicating or not.

Nearly everyone is familiar with various examples of more advanced forms of undoubted communication among nonhuman animals that you might describe as bordering on language. For instance, researchers have pointed to chirping in prairie dogs as a form of language, to alarm calls in vervet monkeys as an example of vocabulary, to whale songs as having syntax, and to the complex chatter of dolphins as language itself. There is even a yearly meeting of scientists, called EVOLANG, which convenes to discuss this subject. Even proponents of the idea that language exists in non-human species recognize, however, that there is much more hypothesizing going on than testing of hypotheses.

Birds have been examined as organisms with exceptional communicative abilities, and indeed they are perhaps the one group of organisms where experimentation in language is the norm and not the exception. The way a young songbird learns a song turns out to be very similar to the way humans learn language. An adult male tutors the young bird learning the song, and in zebra finches there are two phases to this learning. The first phase is memorization, while the second is a "matching" phase in which the learning bird attempts to match its own vocal output with the memorized song used as a template. The template is somewhat rigid, and the songbirds seem to be pretty hardwired to deal with them. This is a bit like one of the two theories that linguists use to explain language acquisition in humans. This theory derives from the work of Noam Chomsky, who suggested that all languages share a grammatical core, or template, and that this reflects an innate disposition to language by humans. The second notion suggests in contrast that, although there are similarities in grammar across human languages, they are far outweighed by the differences. This way of thinking about language simply proposes that, because there is so much diversity in how languages work, there is no single preconditioned innateness for language in humans. Instead, in the words of scientist Johan Bolhuis, the "general purpose learning capacity of the human brain is sufficient to explain language learning."

Nonetheless, the similarities are striking. For instance, babbling in human infants is seen as a critical part of language learning. Deaf children will babble with their hands as a prelude to sign language. Some songbirds twitter or produce subsongs, and some researchers consider this behavior a form of babbling. In addition, other stages of language learning in humans are apparently matched in songbirds and are considered to parallel the steps humans take toward full-fledged language. Bolhuis and his colleagues have examined such parallels and remark that "language development and birdsong learning have parallels at the behavioral, neural and genetic levels." The networks of synaptic connections in the all-important cortical regions and in the basal ganglia are well worked out in humans, and it appears that similarly constructed neural networks exist in songbirds, too.

In fact, though, it appears that this wiring is actually missing in birds that are considered more basal in the bird phylogeny. What this means is that these more basal species diverged from the common ancestral types of birds earlier in evolutionary time than the more very highly derived songbirds did. Thus, the more complex neural connections of the brains of songbirds and parrots must have evolved convergently with ours. If the neural connections were indeed the factor that tips us off to the presence of language, then the "common wiring" of human and bird brains would suggest that birds do have a form of "language." But whatever it is, it is not homologous with human language.

When it comes to closer relatives such as bonobos and chimpanzees, we may more fairly claim to be seeing communication systems containing some of the elements at least that homologously underpin our own complex vocal communication (which of course also includes visual components such as gestures and body language). But there is a fundamental difference between the way in which apes deal with auditory and visual symbols and our own way of processing such cues. Some apes have learned to recognize a fair vocabulary of spoken words (though the nonhuman record is in fact held by a sheepdog) and to respond to them either by carrying out verbal commands or by hitting appropriate visual symbols on a computer screen. But the apes' use of such symbols is inherently limited, and they cannot use them to produce structured statements about the world as we do.

The conclusion is that, although the apes' cognitive abilities are impressive (and in fact, complex spontaneous behaviors observed in the wild, such as the use of stones to crack palm nuts on "anvils" or termite "fishing" with stripped twigs or bushbaby "spearing" with sharpened branches are to us at least as demonstrative of cognitive complexity as any responses set up in laboratory situations), apes are simply not processing information in their brains in the ordered way in which we do it. So, even though bonobos may indeed be giving us an indirect glimpse of a cognitive "stage" through which our remote ancestors might have passed, what we see in them is at most an echo of the foundations on which our own peculiar form of consciousness is based.

In the end, then, looking at putative "languages" in other species is important mainly in providing the wider context within which we need to understand how our own consciousness emerged. And in this context we need to be particularly careful not to mix metaphors with history. The most likely solution to the questions raised by the huge number of observations out there on nonhuman cognition and "language" most likely won't be a single solution at all but rather a bunch of bits of information that shed light on the process of how we acquired language and our unique cognitive style. In the process we will certainly find out how little we currently know and, let's hope, come to understand just how much we need to find out.

Singing Ancestors and the School of Rock

In the 2004 movie *School of Rock*, Jack Black plays a down-and-out rock-and-roll musician who poses as a grade-school music teacher to make ends meet. As an impostor instructor, his character teaches the kids in the school, among other things, about the history of rock and roll. He uses a branching diagram to detail the evolution of rock and roll from jazz to the music of the 1980s (apparently for Mr. Black, rock music stopped evolving with the Dead Kennedys and the Meat Puppets). Another popular musical history lesson is the *Evolution of Dance*, a YouTube video by comedian Jason Laipply that has been viewed nearly 200 million times online. Laipply starts his history even later than jazz, with the swinging hips of Elvis, and culminates the evolution with 'N Sync. But while Black's diagram is treelike, Laipply's presentation, though incredibly entertaining, is definitely scala naturae at its worst. And in both cases the comedians get in on the act pretty late. Music actually dates back a whole lot farther than either of them thinks, much farther than Elvis Presley or Lionel Hampton—as hard as this may be to admit for anyone who loves rock and roll (our apologies of course to Mozart, Beethoven, Puccini, Stravinsky, Copland, or for that matter the composers of the Gregorian chants).

There can be no doubt that music as we know it is the invention of our species. The first musical instruments that can be safely called such are

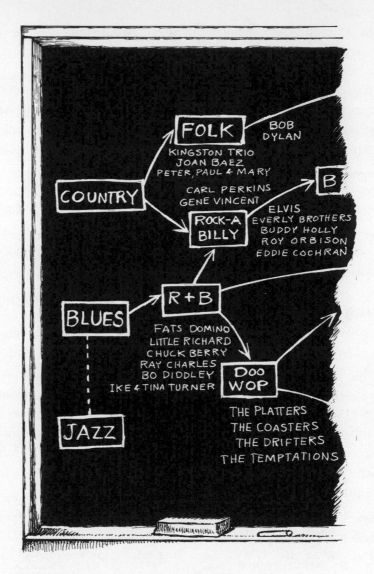

The evolution of rock and roll from *School of Rock*.

about thirty-five thousand years old. And although there are some controversial artifacts from Neanderthal sites, to date no reliably identified musical instruments have come from any deposits older than thirty-five thousand years, and are all are associated with *Homo sapiens*. Despite this, some think that Neanderthals might have had musical abilities (see Steven

Mithen's *Singing Neanderthals*), but that is a story that awaits more research. For now we can be pretty confident that music was a product of the kind of brain that is unique to *Homo sapiens*.

How music is processed by our brains is as complicated as other auditory-based sensations like language. The gateway that music goes through is, after all, the same portal that spoken language uses: the ear. In the AMNH exhibition *Brain: The Inside Story*, the designers wanted to demonstrate the uniqueness of our brain on music and chose to use the functional magnetic resonance imaging work of the exhibition's co-curator, Joy Hirsch. Joy is a neurobiologist at Columbia University and directs the Program for Imaging and Cognitive Sciences, a brain-imaging facility at the medical school. She and her students scanned the brains of Yo-Yo Ma, the famous cellist, and a funk or acid jazz musician from Venezuela, Cheo (José Luis Pardo), of the group Los Amigos Invisibles. Hirsch scanned their brains while they were watching and listening to films of themselves playing music. The scans of both brains are extremely interesting. As you watch the scans of Ma's and Cheo's brains, you see large swaths of red and yellow bathe the active parts of their brains stimulated by viewing themselves. It is no surprise that the areas of their brain where sound is processed glow red and yellow. But oddly enough, as they listen and watch themselves play music, their motor areas are also going wild. Some scientists think that this response might be the result of what are called mirror neurons firing.

Mirror neurons were first discovered in monkeys by Giacomo Rizzolatti, Leonardo Fogassi, and Vittorio Gallese. These researchers were training monkeys to reach for bananas and other objects while they recorded the signals from single neurons to characterize the role those neurons played in muscular movement. After a lot of training of the monkeys, and after obtaining strong data that the neurons they were working on did indeed control movement, one of the researchers serendipitously reached for a banana while the neuron was being assayed, and as a monkey watched him. And the neuron that was being tested went wild, indicating that the watching monkey's movement neuron was firing simply as a result of watching someone else reach for a banana.

Mirror neurons have now been found in a lot of organisms, and it was not too surprising that Ma's and Cheo's brains lit up in these areas as if they were actually playing the instruments. But when deeper areas of the two musicians' brains were scanned, a big difference began to be evident. While he listened to himself, Cheo's brain lit up in areas responsible for self-image, and he was extremely active in the areas of his brain responsible for creativity. Ma's brain, on the other hand, lit up in so many places that it was hard to tell which parts of the brain were most important. And although we are sure that Cheo's brain was working pretty hard, Hirsch called Yo-Yo Ma's brain the "most beautiful brain" she had ever seen, just because of the complexity of the areas lighting up. Hirsch was so impressed because of the richness of the neural highways connecting all of those areas that are musically active in Ma's brain.

But what is it about music that makes it so appealing to our brains? Part of the answer involves solving the chicken-and-egg problem of language and music. To some working on the problem, the question of which came first—language or music—is very important. And like many of the other questions we have discussed with respect to the so-called higher cognitive abilities, this one is much disputed. Cognitive scientist Steven Pinker suggests that music was, and is, simply not that important for the survival of our species. He calls it "evolutionary cheesecake." In essence, this means that it's nice and sweet like cheesecake but was not a part of the diet our ancestors needed to survive. This cheesecake idea implies that music came *after* language. Language in this view is equivalent to sugar and protein in the diet, while music is the cheesecake that came later as a result of liking sugar and protein. Others such as Daniel Levitin disagree. Levitin sees music as an important agent of sexual selection in human populations. Perhaps knowing for sure whether music is somehow adaptive will help resolve this question, but maybe it is the wrong question to ask. What is most likely, in fact, is that both music and language are emergent products of a more generalized underlying capacity in the brain that underwrites them both. We will discuss the evolution of that capacity at length in chapter 10.

9 DECISIONS, BEHAVIORS, AND BELIEFS

TOWER OF HANOI

Insight into how the physical workings of the brain produce decisions, knowing how and why we have spirituality and depression, and explaining bizarre behaviors such as hallucinations are important aspects of understanding our brain's evolution. One helpful avenue toward clarifying these matters is called neuroeconomics. This approach, looking at situations in which advantages and disadvantages are relatively easy to weigh, may well in the longer term provide us with a portal through which we may begin to understand the neural underpinnings of more personal decisions in which the parameters are much less clear-cut. Right now, though, we still have only a superficial understanding how our brain translates all of those inputs into what we consciously (or, indeed, unconsciously) experience as the workings of our minds. We are well on the way to assembling all the many pieces; but we have yet to penetrate exactly how they function as an ensemble. In other words, how brain translates into mind remains a mystery. So let's take a look at some of these higher cognitive functions to see where neuroeconomics can lead us.

When You Get to the Fork in the Road . . .
Take It (Yogi Berra)

As we explained in chapter 5, we make many of the decisions about our lives with the inner parts of our brains. This is because our brains evolved to deal with incoming information in as efficient and rapid a fashion as possible. Our brains may not be optimized to specific tasks as a machine can be, but they are still operating under all the same constraints that have governed brain evolution in other organisms. The convergent similarities between our brain regions and how they are wired, and their counterparts in some quite remotely related forms, should be evidence enough of how influential these constraints have been.

How do we make decisions that don't need rapid solutions? And what happens when we are confronted with conflicting information? In the *Brain* exhibition at the American Museum we address two aspects of processing information from the outer world that our prefrontal cortex has contrived to take over.

The first concerns conflicting information. A Stroop Test makes the point that sometimes we have to deal with conflicting information sets. This test consists of two columns of words describing colors. The words themselves are colored, and in the left column the letters making up the words are colored to coincide with their meanings. In the right column the words do not coincide with their colors. For instance, in the left column the word "red" is colored red, and the word "yellow" is colored yellow. But in the right column the word "red" might be colored blue, and the word "yellow" might be colored green. The subject of the test is asked to yell out the color of the words in the left column (not what the word says but the color of the word). Usually the subject will get through this list rapidly and with no mistakes. But the list on the right always slows the subject up. It takes almost twice as long to get through the list, and even then mistakes will most likely be made. As the exhibit label states:

Every day, your brain swims with sensations, thoughts, and memories that compete for your attention. To make your way through the world, you need to ignore some things and focus on only a few. The Stroop Test presented here tests your ability to do just that. The way this works in your brain is described below:

Step 1 is actually reading the word: Brain areas devoted to vision and language comprehension (Wernicke's area) help you read the word.

Step 2 is naming the color: Brain areas used in vision and language production (Broca's area) help you recognize and name the color of the letters.

Step 3 is detecting conflict: The cingulate cortex detects the conflict between the word and the color, then relays the message to the front of the brain.

Step 4 is controlling your attention to focus on the task at hand: Clusters of neurons in the prefrontal cortex help you focus your attention on the color alone.

When you resist the urge to read the words, you are using the executive centers at the front of your brain, along with a deeper part of the cortex that warns you to stay in control.

So it is our prefrontal cortex that helps us slow things down a bit and correct the information.

Another aspect of how our prefrontal cortex helps us with decision-making is forward planning. Again, the *Brain* exhibition has a very nice way to explain this. The interactive used is called a Hanoi Tower, and it consists of a set of three sticks, with discs on the stick on the left. The three discs go from largest on the bottom, through medium, to smallest on top. The object of the game is to move the three discs from the stick on the left to the one on the right, with the discs in the same order as at the start of the game and taking the fewest number of steps. The rules of the game stipulate that you cannot move more than one disc at a time and that you cannot put a larger disc on top of a smaller one. With three discs there is only one way to do this in the minimal number of steps, which is seven. We guarantee

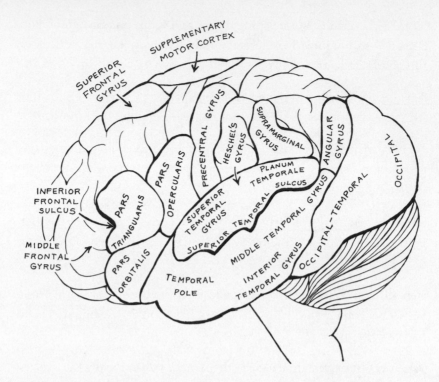

Decision-making regions of the brain.

that if you do not plan each move ahead, you will not hit the minimal number of moves—and that even when you plan ahead, you sometimes won't get it. The writers of the *Brain* exhibition explain this planning procedure as follows:

> When you plan the moves for the Hanoi Tower game, neurons near the front of your brain come alive. They take in a flurry of signals from the rest of the brain, help you weigh your options and choose the best course—then transmit the plan to your brain's motor areas for action. These major areas of the brain are involved:

> Step 1 is thinking spatially: Visual and spatial processing areas help you see the game pieces and their spatial relationships, and imagine your moves. This thinking is done in the visual and parietal cortex.

Step 2 is thinking logically: Planning areas help coordinate information from other brain areas, use short-term memory to keep track of your thoughts, and weigh the pros and cons of possible moves. The planning areas are located in the prefrontal cortex.

Step 3 is controlling movement: Motor areas prepare the muscles in your fingers and hands to move the blocks around. The motor areas are located in the premotor and motor cortex of the brain.

Okay, hotshot! So you figured out how to do three discs—now how about four? The minimum number of steps for four discs is fifteen, and the problem of planning gets harder and harder the more discs are added to the game. Planning ahead will help a lot, and it is a trait that humans have figured out how to do relatively well—at least, some adult humans.

Decision-making has been an important aspect of our species' success. We make numerous decisions each day, including every time we pull money out of our pockets. Let's illustrate this by returning to neuroeconomics. There are certain things that are necessary for an economic system to work. At a recent talk at the AMNH, New York University neuroeconomist Paul Glimcher, one of the founders of this field, explained that economic choices can be either rational or irrational. He then played a game with us, taking the role of the consumer in a rational system. In this system he preferred apples over pears, pears over bananas, and apples over bananas. This can be symbolized as A>P, P>B, and A>B. The object of the game is to start with an unfavored fruit and end up with the fruit you like most. Each time you have a fruit, you trade for a preferred fruit, paying one dollar. As sellers, we could engage Glimcher in a transaction where he starts with a pear that he wants to trade for an apple, at the cost of one dollar. Transaction over, we both go away happy. We earned a dollar and got a banana to sell later, while he got his apple for only a dollar. Even if he made a mistake and traded twice, he still would have made out pretty well by paying two dollars and trading a pear and a banana for his beloved apple.

Next, he let us be the consumer. But this time he created an irrational choice system. In this system we prefer bananas over pears, pears over apples, and apples over bananas, or B>P, P>A and B<A. As before, all trades for fruits cost one dollar. Glimcher then gave us a banana. We knew, though, that we liked apples better, so we gave him a dollar and traded for the apple. Next, we noticed that he also had pears; since we like them better than apples, we traded him for the pear, and paid another dollar. It's getting expensive, but we are getting what we like most, so we figure it is all okay. But then we realized we liked bananas more than pears, and our choice system compelled us to trade for the banana and pay another dollar. It is at this point that we stopped and smelled the fruit, so to speak. Realizing that we had just paid three dollars to get back our original banana was embarrassing, especially since it was in front of an audience. (After his talk we made sure Glimcher hadn't taken our wallets, too.)

Other aspects of economics also need to be in place for a rational system to work, and these all seem to be compatible with the way our brains have evolved. Much of the decision-making in neuroeconomics focuses on the mechanistic aspects of choice, such as the rational economic system we just discussed. Glimcher uses macaque monkeys to determine what the parameters are, but other researchers have chosen to use capuchin monkeys to approach the basics of rational choice. These researchers can train (trick?) their subject monkeys to use some sort of currency to test their abilities and approaches to making economic decisions. They can also use techniques allowing them to measure the neurological response of the monkeys when making the choices and can pinpoint the brain regions and clusters of neurons involved in the choice process. In the case of the capuchins, the currency is a set of rings; for macaques, it's fruit juice. In a clever experiment, the macaques were trained either to trade juice for a peek at a picture or to demand juice to look at pictures they didn't like—a sort of monkey "pay per view," as Duke University researcher Michael Platt explains. Macaques are hierarchical in their social behavior, and the most dominant male monkeys are respected by their juniors. Females, on the other hand, are not normally a part of this hierarchy.

When macaques trained in using the juice currency are presented with pictures of dominant males, they will pay to see more of the pictures. When they are presented with lower-and lower-ranking males they pay less and less, until they reach a point where they demand to be paid to look at the lowest-ranking ones. When they are presented with the faces of females, they also pay very little or ask to be paid to view them. Now, these are males who are reproductively mature, and anyone who has been to a macaque colony knows that sex is happening everywhere. So does this mean that these males are not interested in sex? Well, as Platt points out, the researchers were showing the tested males the wrong end of the female. Once the males were shown pictures of the female perinea (a polite word for their reproductive regions), the males start anteing up to look at the pictures; a kind of monkey Playboy channel. In fact, the monkeys would pay much more for a peek at a perineum than for a look at a dominant male's face. Not only is the behavioral result interesting, but Platt can also trace the decision-making neural activity of the monkey's brain to specific neurons. And it turns out that there is a close correlation between preference-forming and decision-making and neurons firing in a region of the brain lying on the parietal cortex—namely, the lateral intraparietal area. This area controls where a monkey will look next, and this is important for interpreting information for how to deal with macaque society and hierarchy.

Some humans also pay to view pretty faces. Platt has conducted similar experiments with students and has found that males will pay half a US penny to view pictures of the faces of attractive females and nothing to view faces they found unattractive (whatever "unattractive" means). But females will pay precisely nothing to view an attractive male face. To our knowledge, the corresponding experiment using pictures of human genitalia has not been done. Little wonder—imagine the paperwork for the Internal Review Board (every university in the country reviews any and all experiments done with live animals and humans)!

Our emotions also sometimes collide with our reasoning mind. A good example of this problem is the Ultimatum Game, in which there are two players. One player is given a hundred dollars and asked to split it with the second player. The first player (A) then makes an offer. The second player (B) judges the offer and either okays or nixes it. If the deal is nixed, no one gets anything. If the deal is okayed, then the money changes hands. The game goes on until a deal is nixed. Most As will make a reasonable deal like fifty-fifty in the first phases of the game. And B almost always accepts those initial deals. But as the game goes on, A will more than likely get greedy and make an offer that B feels is unfair. The deal will then be nixed by B, and the game ends with no more exchange of money. The golden goose has been slaughtered. Two things are going on here, with respect to emotion and to logic. In this game, the most logical thing for A to do is to offer as little as possible, but he or she usually doesn't because of emotional issues involving some notion of fairness. B logically should take any offer, because otherwise the game ends and there is no more chance of money changing hands. But B usually will nix the deal if it seems too unfair. Again, emotion and a sense of fairness step in. The emotions are a huge part of our makeup as social animals, and the decision-making process here is complicated by social and emotional factors, for it would be most logical for B to play the game ad infinitum, even if the ratio of the deal were ninety-nine to one against.

More recent data from this game, in which the players are autistic adults, are illuminating in this context of decision-making and social appropriateness. Autistic adult players of the game will both take any offer and offer the worst splits they can get away with. They do this apparently because they cannot assess what the other player might believe, feel, or even do. Finally, functional magnetic resonance imaging has been used to localize the decision-making parts of the brain involved in the Ultimatum Game. When player B is offered an "unfair" deal, one brain area activated is the insular cortex, a region associated with the emotion of disgust. Another is the anterior cingulate cortex, a region associated with difficult decision-making. Apparently, when you are offered a raw deal, your insular cortex responds with disgust. After that, your anterior cingulate cortex kicks in to try to resolve whether disgust or no money is worse. If disgust is judged to be worse, then you reject the deal. If poverty is judged to be worse, you take the deal, but the bad taste in your mouth remains. The behavioral part of a similar experiment has been tried with capuchin monkeys, and they respond in much the same way as normal human beings, as do dogs.

In such experiments as these, neuroeconomics has already aided in our understanding of behavioral features such as reward acquisition, gratification delay, strategy development, and cooperation. But beyond its role in understanding how the brain works and in clarifying regularities in mental function, Paul Glimcher also hopes that neuroeconomics can help unify the science of economics itself. As he points out, there are two theoretical approaches to economics. The first involves prescriptive methodologies that aim to develop mathematically efficient algorithms for discovering efficient and optimal decision-making strategies. The second approach is descriptive, aiming to use empirical data in modeling or describing how people behave in an economic context. Glimcher summarizes his views on the future potential of neuroeconomics, and of allied brain research on economic questions, as follows:

> One recent trend in economic thought that may reconcile this ongoing tension between prescriptive and descriptive approaches is the growing interest amongst both economists and neuroscientists in the physical mechanisms by which decisions are made within the human brain. These *neuroeconomic* scholars argue that study of the physical architecture for decision making will reveal the underlying computations that the brain performs during economic behavior. If this is true, then economic and neurobiological studies which seek to bridge the gap between these two fields may succeed in providing a methodology for reconciling prescriptive and descriptive economics.

As Glimcher's words imply, we haven't got there yet, but how we make economic decisions is the result of one of those "beautiful experiments" that the evolutionary process has done with our brains. And it is fortuitous that we now have neuroeconomics as a tool to help tease apart the steps in the beautiful experiment.

God Helmets, Alien Abductions, and the Blues?

If you believe all the fuss, the Virgin Mary, mother of Jesus, must have had as many fashion looks as Lady Gaga. Almost every year, her likeness is reported to appear on anything from the walls of freeway exits to the surfaces of sandwiches. In fact, just about a decade ago, a piece of grilled cheese sandwich with her image on it was auctioned off for twenty-eight thousand dollars. And this hasn't just been going on recently. Apparitions of the Virgin Mary have been reported ever since the Assumption. Countless and varied images of the event, and of the person herself, have been created by artists. And when you throw in the toast, the water markings on walls, and assorted other apparitions, she appears in an amazing assortment of guises.

Even as diners are finding images of the Virgin Mary on their sandwiches, reports regularly come in from all over the globe from people who have been abducted by aliens. Wild scenarios abound of individuals who have "gone for the ride" or been "anally probed" by such beings. And although the details differ, a lot of the descriptions of such abductions involve the victims' being in an operating theater, completely helpless as some presence examines them, sometimes not so gently and in not such nice places. There are other commonalities, too, such as the aliens' typically large eyes and the lights and saucer shapes of their flying machines.

We are not aware of a scientific literature devoted to the question of how and why some people see the Virgin in markings that others would regard as purely random. But believe it or not, scientists have looked into the matter of alien abductions. And it emerges that sleep paralysis, an interesting phenomenon that regularly occurs in peoples' sleep patterns, might be responsible. This idea makes a lot of sense: most sensations of abduction occur at night, and some people even report being awakened from deep sleep to find themselves seemingly being taken away. Sleep paralysis involves a loss of muscle tone and control, and it occurs during the REM (rapid eye movement) phase of sleep, when the brain (especially the amygdala) is unusually active and the most emotionally fraught dreams are experienced. Indeed, some scientists believe that the function of sleep paralysis is to inhibit the thrashing around, and possible injury, that might occur were the sleeper to act out those dreams.

Because of its association with powerful dreaming, many kinds of recalled experience are connected with sleep paralysis, and indeed some recovered memories of sexual abuse have been attributed to it. The key feature of sleep paralysis is the sensation that one is powerless, unable to move, and it is often accompanied by hallucinations that someone or something is nearby. When there is a slight suggestion of aliens or the person concerned is susceptible to hypnotic effects, sleep paralysis, which may linger as the individual awakes, will often be interpreted in retrospect as an alien intrusion. David Forest at Columbia University Medical School adds this about subjects he has studied claiming alien abductions:

> I noted that many of the frequently reported particulars of the abduction experience bear more than a passing resemblance to medical-surgical procedures and propose that experience with these may also be contributory. There is the altered state of consciousness, uniformly colored figures with prominent eyes, in a high-tech room under a round bright saucerlike object; there is nakedness, pain and a loss of control while the body's boundaries are being probed; and yet the figures are thought benevolent.

Forest's work sheds light on the power of suggestion in the generation of episodes of supposed alien abduction, and perhaps will also be useful in getting to the root of some recovered memories of sexual abuse.

We hesitate to mention more generalized feelings of spirituality under the same rubric as alien abductions. But like nearly all our other experiences, spiritual feelings are filtered through the brain and are thus at least in principle accessible to neuroscience. In one famous case, spiritual experiences were induced using electromagnetic stimulation of the temporal lobes. The Nobel laureate Francis Crick once described our mental processes as the result of a "vast assembly of nerve cells and their associated molecules"—in which case, outside forces like a magnetic field should be able to stimulate these cells, or alter the activity of the molecules, to generate thoughts in our brains. To test this notion, Michael Persinger, a psychologist at Laurentian University, devised a helmet that was wired to send out a relatively strong magnetic force. He used this helmet to stimulate more than a thousand subjects, some of them hard-bitten journalists. An electromagnetic field was focused on various regions of the brain while the subject was observed and asked a battery of psychological questions.

Perhaps Persinger's most intractable subject was Richard Dawkins, author of *The God Delusion* and general purveyor of bad news about religion. Dawkins once said, "Religion is about turning untested belief into unshakable truth through the power of institutions and the passage of time." And that is just one of his more benign statements on the matter. Dawkins was perhaps not the ideal subject for this experiment; and, perhaps predictably, other than experiencing some dizziness he was completely unresponsive to the test. But Persinger claims that nearly 80 percent of all of his subjects responded with feelings related to religion and spirituality, some of them undergoing full-blown rapturous experiences. And while his "god-helmet" work has been criticized by other scientists, it is undeniable that magnetic fields have an impact on how our brains function.

The first transcranial magnetic stimulation (TMS) experiments were done in the mid-1980s and involved simple stimulation of the motor cortex and the consequent muscular contractions. The technique has since been adopted by neurobiologists in a variety of clinical applications, most commonly to help diagnose damage in patients with strokes or other brain damage. More recently, TMS has been thought to be useful in the treatment of depression. However, the US Food and Drug Administration (FDA) found that there was no proof that TMS was efficacious in this role, and clearance for marketing such devices was denied in 2008—although the device was placed on the market a year later through a loophole in FDA rules. Part of the FDA's problem with using TMS to treat depression was that the control groups in trials could guess if they were being treated by TMS, so a placebo effect was more likely to occur. But whatever the actual impact of TMS on depression, it is interesting, and perhaps even significant, that the stimulation of a small part of the brain could have such an impact on such complex human behaviors as spirituality and depression.

Persinger's more recent work focuses on meditation and the brain and on perceived time travel generated in the brain, perhaps because these are much less provocative and controversial areas of research.

It is early days yet, so at present we can be confident only that the human brain's generation of our consciousness and its various products is somehow determined by what is going on inside it, sometimes in very limited functional areas. This is clearly true for alien abductions and visions of the Virgin. But (however infectious they may be) such specific perceptions are not the rule, while spirituality in the general sense is a more or less ubiquitous feature of human populations, and thus something it's even more important for us to understand. And the obvious starting question is why it exists in the first place.

One notion is that spirituality is adaptive for social behavior in humans. The idea here is that groups of people needed a unifying context for their existence and survival, and religion, ceremony, and rituals offer a good way to attain group cohesion. Another argument is that religion soothes the continually nagging and existentially harrowing knowledge of our mortality. Some even believe the "God gene" proposition that specific genes have been involved in the evolution of religion in humans. Still, whenever a claim is made that "the gene" for a complex behavior has been found, a red flag should go up. Alternatively, we suspect that religion (like the invention of beer) may well have arisen independently in disparate societies and cultures, based on the innate propensities of the unique form of human consciousness we'll discuss in chapter 10. Once it had become possible to imagine the existence of other worlds, of other planes of being, it became inevitable that they *would* be conceived of in human societies scattered across the globe, albeit quite plausibly at different times. As in the case of beer, the ingredients were different each time religion was invented, but the overall principle remained the same. So if any particular genes *were* involved, they probably weren't the same genes from one ancient culture to the next. Much more likely, though, spiritual feelings are a product of the more generalized capacity to think in symbols, in which case religion, while seemingly a "universal" human behavior, is probably best viewed as a complex set of behaviors sharing a substrate but independently derived in different cultures.

And Then There Are Teenagers

Most parents know, or will soon find out, what it is like dealing with teenage offspring. Although the experience is often rewarding, it is also complicated by the fact that the path to twenty-three is paved with decision-making on the part of the typical teen that can most charitably be described as "bad." The teenage brain is also one of the best examples we can think of to illustrate the "messiness" of our brains that we spoke of in chapter 1. Almost certainly, any parent designing a brain would devise some way of skipping the teenage years. This is because, as the brain develops through the teen stage, the parietal and temporal lobes and all of the associated structures are the first part of the cortex to mature, which means that the sensory, auditory, and language centers are fully developed relatively early. This is why parents are often astounded by, and proud of, the development of their teens' abilities in speaking, intellectualizing, and athletics. So far, so good. But the problem lies with the prefrontal cortex region, which is a more recent evolutionary acquisition and is much slower to develop. We've seen how the prefrontal cortex is involved in thinking, planning, and some decision-making. Like anyone else, teens need to make decisions. And instead of relying on their incompletely developed prefrontal cortexes, they fall back on the next best and more developed structures, their amygdalas. As we saw in chapter 5, in vertebrates this region of the brain is largely involved in emotional response. This is one reason why teens so often make really bad decisions and are incredibly impulsive.

Some experimental evidence exists to support this interpretation of teen behavior. When adults are shown the face of a person who is afraid, nearly 100 percent of them identify the emotion in the picture correctly. But when teens are shown the same pictures, nearly 50 percent get it wrong. Even more telling are imaging data that point to the prefrontal cortex as the region of the brain that is mediating this interpretation in adults. When teenagers are shown the picture in a functional magnetic resonance imaging experiment, their amygdala shows unusual activity—as do other regions of the brain that are not usually associated with assessing emotions. So, even though we really don't like making analogies between brains and computers, we completely understand the description of the teen brain as an incredibly advanced supercomputer—before the operating system and software have been installed. Again, for the sake of parents, and of the teens themselves, a rationally designed brain would have short-circuited this developmental stage.

So why do we have this developmental discrepancy? Some evolutionary psychologists suggest that teenage years are adaptive—that is, that they are there because of natural selection. In other words, the bizarre behavioral aspects of the teenage years offered some increase of fitness in groups at a pre-settled stage of human existence. It's hard to see just what that advantage would have been, and it seems to us that this "unbalanced" phase of brain development is more plausibly traced to the general embryological pattern that was noticed by Ernst Haeckel back in the nineteenth century: that the more ancient structures seem to occur earlier in development. In this context, the architectonic refinements of the prefrontal cortex, of which we'll read more later, were the latest in the long series of acquisitions that led to the human brain as we know it today—and are thus the latest to develop and mature. But of course, almost any explanation will do, if it keeps us from committing our teenage offspring to asylums until they are over it.

10 THE HUMAN BRAIN AND COGNITIVE EVOLUTION

CAVE ART

By now you should have a decent general picture of the brains of other organisms and of what brains in general do. In the previous chapters we have laid the groundwork necessary for a return to the task we laid out in Chapter 1: thinking about thinking. We are finally ready to tackle the question of how, in particular, the human brain evolved and how its special capacities were acquired. To approach this central undertaking it may help to summarize the intricacies of the human brain, so we begin with a brief recap of its structure.

The human brain consists of an amalgam of parts that are both structurally and topographically distinct, each one acquired at a different stage of the long human evolutionary journey. Deep within the brain are the ancient structures of the brain stem, connecting it with the spinal cord that is the conduit to the rest of the body and controlling such basic functions as breathing. Behind this complex is the bulging and wrinkly cerebellum that is concerned with balance and movement and mediates higher functions as well. And above the brain stem lie the various structures loosely grouped as the limbic system. Such structures include the amygdala, which processes such emotions as fear and gratification; the hippocampus, which is involved in the formation of long-term memories and in the ability to find one's way around; the cingulate gyrus, which helps regulate autonomic functions such as heart rate and blood pressure, as well as attention; and the hypothalamus, which secretes hormones important in regulating the autonomic nervous system as a whole, including other limbic structures such as the aforementioned cingulate gyrus. The hypothalamus is vital in stimulating such functions as hunger, thirst, sleeping and wakefulness, and sexual activity. That one brain area may carry out a variety of different functions suggests many of them may have been coopted for new uses at various points in evolutionary time, which is actually a pretty routine piece of evolutionary jury-rigging.

Atop the limbic structures is the cerebrum, which consists of the basal ganglia below and the cerebral cortex above. The basal ganglia are involved in motor control and learning, whereas in coordination with the cerebellum the cortex controls all the voluntary actions of the body, including higher cognitive functions. The cortex is a thin sheet of neural cells, only a fraction of a centimeter thick, that has so greatly expanded during mammalian evolution that in bigger-brained species it has had to be creased and folded to fit into the confines of the cranial vault, much as you can scrunch up an item of clothing to fit it into the corner of a bulging suitcase.

Traditionally, the major creases on the outside of the brain that are produced by this scrunching process have been used to identify major functional areas of the cortex. The cerebrum itself is bilaterally symmetrical and divided into left and right halves, each half concerned not only with the opposite side of the body but at least in part with different functions: the right hemisphere is generally identified with creativity and the generation of new ideas, whereas the left is more rigorously logical. The two hemispheres "talk" to each other via a big bundle of fibers known as the corpus callosum. Topographically, four major cortical "lobes" are represented on each side of the brain, each one distinguished by major creases known as sulci. The parietal lobes lie at the upper sides and are concerned with voluntary movement, recognition of stimuli, and orientation. The occipital lobes lie at the back and are involved in visual processing. The temporal lobes are positioned below the parietal cortex and are devoted to auditory functions, memory, and speech. Finally, farthest forward are found the frontal lobes, in which reside reasoning, planning, problem solving, the control of emotions, and a host of other so-called executive functions. Within each major lobe, subdivisions are readily recognized, and delimited by deep wrinkles. The primary motor cortex, for example, is a vertical strip of cortex, lying at the back of the frontal lobe, which works in conjunction with a "premotor" area just in front of it to plan and execute movements of the body, based on information supplied by other brain areas. Adjoining the premotor area is the "prefrontal" division of the frontal cortex, the region above the eyes that plays a vital role in complex cognition. Its purview includes such key areas of mental activity as decision-making, the expression of personality traits, and the moderation of social behaviors. Similarly, the small patch within the left frontal lobe known as Broca's area plays a key role in motor functions that include control of the speech production apparatus, and the equally limited Wernicke's area, long associated with the comprehension of language, is located at the bottom of the parietal lobe.

As the brain grows within the skull during the development of each individual, an intimate association develops between the outer form of the brain and the inner contours of the cranial vault that contains it. All that lies between the two are three thin layers of membrane and some blood vessels. As a result, the internal morphology of a braincase constitutes a remarkably accurate record of the external form of the brain that lies or lay within: lobes, wrinkles, furrows, sulci, and all. Of course, there is a limit to what you can tell from simply looking at the outside of a brain; it is what's going on inside that is critical, and imaging studies in living individuals have shown that many actions are more diffusely distributed in the brain than a purely modular division into externally visible functional areas would suggest. What's more, detailed studies of particular regions of the brain suggest that the various parts of it may differ quite strongly among mammals, including primates, in their internal architecture.

Studies have begun to show, for instance, that even though the vastly expanded human brain does not exhibit any readily identifiable discrete structures that you can't also find in apes, the neuronal architecture of individual brain regions may be quite distinctive. For example, let's look at a part of the prefrontal cortex that lies right at the front of the brain, above the eyes. This cortical region, known as Brodmann Area 10 (BA 10), has long been believed to play an essential role in complex thought. And now it turns out not only that BA 10 is somewhat expanded in humans relative to apes but also that the neurons within it are more loosely packed in humans. This arrangement allows each neuron to send out more of those branching axons and dendrites, and thus to develop a more complex set of connections with its neighbors. In humans, BA 10 is also particularly rich in what are known as Von Economo neurons, nerve cells that are thought to be specialized to facilitate high-speed transmission of information from one part of the brain to another. It is perhaps significant that differences such as these between humans and apes are not—or have not yet been—found in other cortical regions such as the visual and motor areas. Putting all this information together not only reinforces the long-held suspicion that the prefrontal cortex is crucially devoted to the integration of data from multiple brain regions that lies at the seat of our human consciousness but also emphasizes that the internal reorganization of the brain was at least as important as overall size expansion in human evolution.

Still, even granted the validity of both of these last points, on a general level both the external form of the brain of a mammal species and its overall size (particularly its size relative to that of the body it has to control) provide potentially important sources of information about its overall cognitive capacities. Knowledge of this kind is, of course, particularly significant in the case of extinct species, in which neither those capacities nor the microarchitecture of the brain itself can be directly observed. As a result, it's fortunate that information about the external brain morphology of fossil species can often be retrieved by making physical "endocasts" of the insides of their braincases, either directly or via new techniques of neuroimaging. Indeed, in some cases nature has already done the job for us: not a few natural endocasts are known in the fossil record. These endocasts were formed when the cranial vaults of dead individuals were filled up by sediments before fossilization. And because endocasts are the only record we have of what the brains of extinct hominid species (primates that were more closely related to us than they were to any of today's great apes) looked like, they have been scrutinized, albeit to rather mixed effect, by the specialized paleoanthropologists who call themselves "paleoneurologists."

The Starting Point

Given the inherent limitations of endocast studies, it's fortunate that those brain imprints are not the only means we have for making inferences about early human cognitive evolution. From the earliest neural scaffoldings, we have been in the fortunate position of being able to employ comparative evidence from living creatures in fleshing out our picture of the evolutionary story of the brain. In the case of the hominid family (or the hominin subfamily, as many prefer—take your pick, it really makes no practical difference), we can use our closest living relatives, the great apes, to help us establish a starting point. After that we're on our own, because there are no living analogues in the world for later stages of human evolution. But since the apes have brains that are pretty close in size (both absolutely, and relative to their body volumes) to those of our most ancient ancestors, the great apes can help us a lot in making inferences about how those ancestors perceived and interacted with the world around them.

The best understood of our close living relatives are the bonobos and chimpanzees of tropical Africa, and the more we study these remarkable creatures as cognitive beings, the more impressive they appear. Hardly a month passes in which chimpanzees are not reported to do something we had previously thought only we humans did. Early on, it was the careful preparation of twigs for use in "fishing" for termites in their mounds; then it was using rocks to bash hard palm nuts on stone anvils to extract their edible flesh (an activity that has even lately been discovered to have an "archaeological record" stretching back several thousand years); now such activities even include the preparation and use of wooden "spears" to impale bushbabies sleeping deep within tree holes. Chimpanzees band together to hunt, capture, and eat colobus monkeys in the trees and young bush pigs on the ground, and they even "conspire" to murder adversaries belonging to other social groups. They are truly amazing animals; and nobody looking into the eyes of a captive chimpanzee in a zoo (or, if they're lucky enough, in the wild) could ever deny that they are seeing a lot of themselves in the ape. But there is still a difference: cognitively as well as physically, they are so close to us but at the same time so far.

After extensive long-term coaching efforts, for example, bonobos and chimpanzees have still failed to grasp the basic principles on which early hominids began to make stone tools, about 2.5 million years ago or perhaps even a lot more. They rapidly grasp the notion of using a sharp stone flake to cut a cord that holds a food reward out of reach, and they have even reached ingenious solutions to the problem of obtaining such flakes; but the principles of striking one piece of rock with another at just the right angle and with the amount of force needed to detach a sharp flake have consistently eluded them. Similarly, great apes have shown in laboratory situations a remarkable ability to recognize visual and verbal symbols, and even to combine them to make statements; but they do not appear able to mix them up to make an infinite number of statements using a finite number of basic elements, as we do every time we use language. To be fair, it must be recognized that these apes lack a vocal tract capable of producing the sounds we employ in speech; but this does not disguise that, while their additive use of symbols (for example, in "take . . . red . . . ball . . . outside") does allow them to understand simple statements and even, with the aid of specially designed computer keyboards and screens, to make them, it severely limits the complexity of those statements and leaves no room at all for abstraction. What's more, all of these behaviors are elaborately elicited in laboratory settings and do not reflect the cognitive responses the apes make in natural conditions. Indeed, although chimpanzees have an impressive repertoire of spontaneous natural vocalizations—even compared to other nonhuman primates—virtually all appear to be associated with, and limited to, immediate emotional states.

For all the many studies that have been carried out on ape cognition, none has revealed any evidence among our living relatives of the modern human penchant for "symbolic" cognition. It seems we modern humans are alone in our ability to dissect the world in our minds into an extensive vocabulary of intangible symbols that we can then combine and recombine to make an endless number of statements, not only about the world as it is, but also as it might be. This symbolic style of intelligence is what makes our way of perceiving and dealing with our surroundings unique in the living world. And to a large extent the study of the evolution of modern human cognition involves the search for evidence of when and how this unique ability was acquired. Clearly, it was not present from the start: it was not primitive for hominids, and the subsequent evolutionary history of our family did not consist of the simple long-term honing of a capacity that had existed in rudimentary form at the beginning. Even if they were set apart by their bipedal proclivities, and though they were doubtless smart by prevailing standards, cognitively speaking the earliest hominids were pretty run-of-the-mill higher primates. They did indeed have brains much larger than any of their known primate contemporaries had, but those brains were no bigger than those of today's great apes, even when body size is taken into account. The very first hominid species known have not yet yielded decent cranial endocasts, but there is little reason to think that their brains would have differed significantly in their external proportions or internal organization from those of, say, chimpanzees. All in all, we have no compelling grounds for believing that our most primitive relatives represented any cognitive advance relative to what we see among our closest relatives living today.

Using these observations, we can reasonably take modern apes as a yardstick by which to gauge at least the general cognitive state of the early hominids. Not long ago the distinguished cognitive psychologist Daniel Povinelli concluded that a fundamental distinction between the ways in which humans and chimpanzees view the world is that, whereas humans form abstract views about other individuals and their motivations, "chimpanzees rely strictly upon observable features of others to forge their social concepts." He continued by surmising that, if this were correct, it "would mean that chimpanzees do not realize that there is more to others than their movements, facial expressions, and habits of behavior. They would not understand that other beings are repositories of private, internal experience." This is all, of course, simply inference from what we can observe of chimpanzee behaviors, both under natural conditions and in the laboratory. We have no way of accessing an individual chimpanzee's subjective state of mind. But it seems a reasonable conclusion to draw, and it would imply as well that individual chimpanzees do not have equivalent awareness of *themselves.* They *experience* the emotions and intuitions stirred in their brains, and as intensely social beings they may express them or inhibit them, as appropriate to the social situation. However, Povinelli believes that they "do not reason about what others think, believe and feel . . . because they do not form such concepts in the first place." And it seems reasonable to conclude that this also applies to self-reflection. For if individual chimpanzees are unable to develop concepts of the internal lives of others, it is highly improbable that they possess equivalent insight into their own interior existences.

Profound as it is, this cognitive difference between apes and modern humans will not always necessarily produce distinctive superficial behaviors; and though it's true that the actions of chimpanzees and human beings sometimes seem astonishingly similar, we should be extra careful not to be misled by such similarities. We can certainly explain why those similarities should be there, by citing our enormously long-shared evolutionary histories, but it remains true that similar observable behaviors may mask hugely different cognitive processes. Still, even though a great deal has happened (on both sides) since human beings shared an ancestor with any ape, most authorities find it reasonable to conclude that cognition of the kind we infer for chimpanzees and other great apes can be used to give us a reasonable approximation of the cognitive point from which our ancestors started some seven million years ago. In Povinelli's words, we may assume that those ancestors were "intelligent, thinking creatures who deftly attend[ed] to and learn[ed] about the regularities that unfold[ed] in the world around them. But . . . they [did] not reason about unobservable things: they [had] no ideas about the 'mind,' no notion of 'causation.'" In the human sense, this implies that they as yet lacked any idea of "self." This we find to be a very plausible characterization of our lineage's cognitive starting point; but limited as it is, it pretty much exhausts what we can usefully say on this subject. To follow the story, we need to turn to our fossil and archaeological records.

Early Hominids

The first evidence we have for beings more closely related to us than they are to any of the great apes comes in the form of a smattering of fossils from central and eastern African sites that date to between around six and seven million years ago. Most important is that all of these fossils have in common (mostly by inference) upright posture, the feature generally believed to have most strongly characterized our earliest ancestors. Everything that followed stemmed from this basic postural shift. Given the miscellaneous assortment of fossils available from this early period, this unusual postural proclivity is inferred from a variety of kinds of evidence. The earliest of the claimed early hominid species is *Sahelanthropus tchadensis*. This species comes from the central African country of Chad, and it is best known from a single skull that was found badly crushed. Its postural proclivities are inferred from the fact, confirmed by a "virtual" reconstruction using sophisticated visualization techniques, that the foramen magnum, the large hole in the skull base through which the spinal cord exits the brain cavity, is rotated to face somewhat downward. This is a tell-tale piece of evidence: in a quadruped, the skull is hung from the front of a more or less horizontal vertebral column, so that the foramen magnum points back, whereas in bipeds such as us, it is delicately balanced atop a vertical spine, so the foramen magnum is tucked underneath and opens down.

In contrast, the 5.8-million-year-old Ethiopian *Ardipithecus kadabba* is known only from a few very scrappy bits and pieces, and evidence for its bipedalism comes from only a single toe bone. This species, however, is believed to be closely related to the much later (4.4 million-year-old) *Ardipithecus ramidus*, of which most of a skeleton was described in 2009 and which carries a signal of upright posture in several bony elements. These include a fragmented skull from which a downwardly turned foramen magnum has been inferred. This *Ardipithecus* certainly wasn't a conventional biped, however: among other bizarre features it had a grasping foot, an adaptation to climbing trees that is otherwise unknown among the earliest hominids. The 6-million-year-old *Orrorin tugenensis* from Kenya is reasonably inferred to have been bipedal from some incomplete leg bones, but it has no associated skull, though some bits of jaws with teeth are known.

All of these putative early hominids also share, to one extent or another, a common configuration of the front teeth involving the reduction of the canines and the modification of the relationship between the upper canine and the anterior lower premolar with which it occludes. In apes the canines are large, especially in males, and the back surface of the upper canine is honed against the front surface of the front lower premolar as the jaw closes. In the hominids, the canines have been reduced in height, and the honing mechanism has been lost, resulting in a change in shape of the tooth rows from a U-shape to a more rounded contour. This change is associated with a reduction in the length of the face. Still, all in all, the earliest hominids make up a pretty assorted lot. Their fossils appear in rocks laid down at a time when the parent continent of Africa was undergoing a climatic trend toward drying out and greater seasonality, a development that led to the fragmentation of forest cover and the expansion of woodlands and woody grasslands. Reconstruction of the habitats of the early hominids suggests that they were exploiting all of the new habitats and is consistent with the idea that they were periodically coming to the ground to exploit the new resources to be found there. Subsequent hominids have been remarkable for their generalist dietary tendencies, and the evidence we have suggests that this was true from the start: whereas modern chimpanzees tend to stick to their preferred diet of soft, fleshy fruit when they leave the forest, *Sahelanthropus* and its like were probably much more adaptable, subsisting on everything available. Certainly, thick molar enamel hints at the processing of much tougher foodstuffs than apes are ever wont to tackle.

Why the early hominids moved around upright when they came to the ground is hugely debated. But the simple reality seems to be that they would never have done so if they had not already been comfortable moving around in the ancestral trees with their trunks held upright. A number of forest-dwelling fossil apes appear to have shown adaptations to uprightness in their body skeletons, and it is not at all unlikely that, under the pressures of the expanding new open environments, more than one ape lineage might have adapted to part-time terrestrial life by moving bipedally on the ground: something that would help explain the relative heterogeneity of the putative earliest hominids, none of which was particularly fleet of foot.

Alternatively, we may be seeing here, at the very beginning, an example of a strong trend that has marked the hominid family ever since—namely, that the story of the hominids has not in the least been one of a single-minded slog from primitiveness to perfection under the benign hand of natural selection. Instead, hominid history has from its inception been one of vigorous evolutionary experimentation, an exploration of the evidently many ways there are to be a hominid. Hominid species have arisen, have gone out to do battle in the ecological arena, and as often as not have become extinct in the process.

Of all the early hominids we know, only two offer us any clues as to what their brains were like. In the absence of a body skeleton, at least that has been described, it's not known what the brain-to-body size ratio of *Sahelanthropus* would have been, but its raw brain volume is estimated at between 320 and 380 cubic centimeters (cc), about the size of a tennis ball. This puts *Sahelanthropus* in line with the average chimpanzee, and it contrasts dramatically with the 1,350 cc average of modern humans (who show a vast range of "normal" brain sizes, from around 900 to 2,000 cc). It's worth noting that within our species, absolute brain size does not correlate well at all with intelligence, however measured: the Nobel Prize–winning French novelist and intellectual Anatole France is said, for example, to have possessed a brain right at the low end of this range. On the other hand, it is highly probable that the average brain size of a *species* is indeed telling us something about how intelligent members of that species were, even if that something is not very specific. And if this is indeed the case, it seems once again that we can look to chimpanzees as a general guide to early hominid smarts. From this perspective, it appears fair enough to say that *Sahelanthropus* (and any other primates, unknown to us at present, that may have been like it) was in all probability the most intelligent creature around in its day; in the absence of any available account of what the brain of this form actually looked like, it is hard to be more specific than that.

Still, what we know of *Sahelanthropus* coincides pretty well with our also rudimentary knowledge of *Ardipithecus ramidus*, which had, if anything, an even smaller brain (estimated at between 300 and 350 cc). In this case, we have a body skeleton to go with the cranial capacity, so we know that the ratio between the two was more or less chimpanzee-like. In addition, the people who described the *Ardipithecus* fossils noted that "the steep orientation of the bone on which the brain stem rests suggests that the base of the *Ar. ramidus* brain might have been more flexed than in apes." They go on to point out that in somewhat later hominids, "a flexed cranial base occurs together with expansion of the posterior parietal cortex, a part of the modern human brain involved in aspects of visual and spatial perception." Perhaps something was already astir in the perceptual realm at this early stage in the evolution of the hominid family.

There's not much else that we can say so far about our earliest precursors from the direct evidence of fossil brains. But there may be something we can infer about them from their ecological context. Descending from the trees to live at least part-time on the ground meant that the primitive bipeds were exposed to a hazardous new environment. The invasion of the forests by more grassy areas entrained the incursion into the early hominids' territory of large numbers of grazing mammals—and of the huge variety of carnivores that preyed on them. Sharing the new habitat would clearly have meant that the slow and relatively small-bodied hominids were exposed to similar new dangers, and this has extensive implications not only for the sizes and organization of the groups in which they lived but for the complexity of their social interaction. It has usually been taken for granted that early hominids would have lived in relatively small social groups, as the basically forest-living apes do. It has, though, been suggested that a more appropriate analogy would be not with our forerunners' closest genealogical relatives but instead with their closest ecological equivalents. And among all living primates it is not the apes, but instead the baboons and macaques, that best fill this bill.

Spending a lot of the time away from the protection afforded by the forests, macaques and baboons are subject to heavy predation, and they accordingly exhibit many of the general characteristics of prey species. For safety, both macaques and baboons live in large groups, carefully arranged spatially so that females and their offspring are in the center, protected by a periphery of reproductively more expendable young males. They are also "refuging" species, seeking the shelter of cliffs and trees at night, when the dangers are greatest. It seems entirely reasonable to expect that the earliest hominids, equally close to a more or less exclusively tree-living ancestry, would have responded to their new environment in a similar way, living and refuging in large groups, within which interindividual interactions would have been both extensive and intense. A social milieu of this kind would have placed a premium on the cognitive ability to comprehend what are known as "higher orders of intentionality" (the ability of one individual to understand that another is aware of a third individual's state of mind, and so on). The larger and more complex the group, the more highly advantageous it is to be able to read and predict what is going on in the minds of other members of the group. Specialists have strenuously argued about the extent to which apes display the rudiments of this ability, which is carried to its zenith by *Homo sapiens* today. But whatever the case, we may perhaps seek the initial stirrings of this remarkable cognitive capacity in the unquestionably complex group living of the earliest hominids.

"Bipedal Apes"

Soon after the time of *Ardipithecus ramidus* we begin to pick up traces of the earliest hominids about whom we can claim to know a substantial amount. These are the "australopiths": members of the genus *Australopithecus*, and forms closely related to it, which flourished in Africa between about 4 and 1 million years ago, as is evident from the family tree we've illustrated. The earliest of them is the species *Australopithecus anamensis*, which is known to have lived in Kenya, and perhaps in nearby Ethiopia, between about 4.2 and 3.9 million years ago. Only fractionally younger than *Ardipithecus ramidus*, this form makes a much better precursor for the later australopiths than does *A. ramidus*. A broken shin-bone, possessing both knee and ankle portions, confirms beyond reasonable doubt that it was an upright biped. What is more, despite some differences, *Australopithecus anamensis* is comfortingly similar in what is known of its teeth and jaws to the next-in-line australopith species, *Australopithecus afarensis*. This is the species that contains "Lucy," probably the most famous fossil of all time.

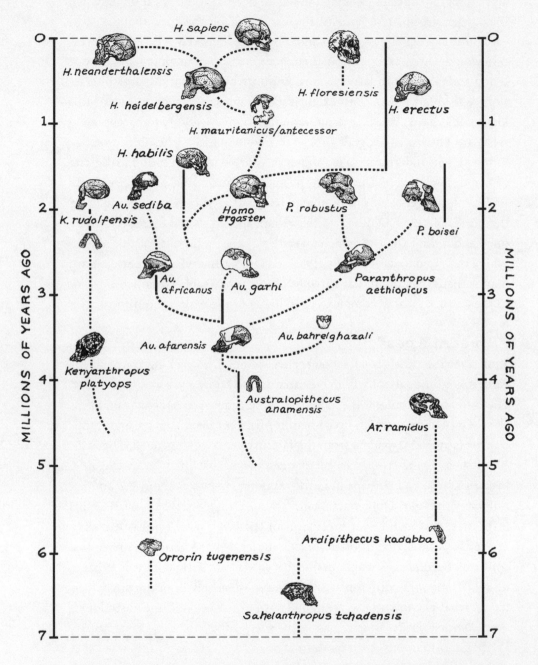

MILLIONS OF YEARS AGO

0 — H. sapiens

H. neanderthalensis

H. heidelbergensis

H. mauritanicus/antecessor

H. floresiensis

H. erectus

1 —

H. habilis

K. rudolfensis

Au. sediba

Homo ergaster

P. robustus

P. boisei

2 —

Au. africanus

Au. garhi

Paranthropus aethiopicus

3 —

Au. afarensis

Au. bahrelghazali

Kenyanthropus platyops

Australopithecus anamensis

4 —

Ar. ramidus

5 —

Ardipithecus kadabba

Orrorin tugenensis

6 —

Sahelanthropus tchadensis

7 —

Tentative hominid family tree, sketching in some possible relationships among species and showing how multiple hominid species have typically coexisted—until the appearance of *Homo sapiens*. Diagram by Jennifer Steffey, © Ian Tattersall.

Lucy is a partial skeleton of a tiny individual who lived about 3.2 million years ago and who in life stood a little over a meter tall. Discovered in Ethiopian deposits laid down by a river that wended through a mosaic of environments ranging from forest through woodland to bushland, Lucy is complemented both there and elsewhere in eastern Africa by literally hundreds of other fossils, ranging from isolated teeth to more or less complete crania, that give us an unparalleled picture of what a hominid species of that remote time was like.

And what a picture! *Australopithecus afarensis* was clearly hominid in its dental features and its bipedalism, no question about that. But it was far indeed from being simply a small-bodied version of us. An upright walker this species may have been, but these ancient hominids were certainly not committed to bipedalism in the way that we are. Lucy and her kind had much shorter legs and wider pelvises than we do, and they would have moved much less efficiently over open terrain despite possessing feet that, if longish, were nonetheless well suited for terrestrial weight bearing. Dramatically gone are the grasping feet of *Ardipithecus*. At the same time, though, these australopiths retained a host of features that would have been extremely useful for moving around in the trees: narrow shoulders with mobile joints, light bodies with long arms, hands bearing long, slender, and curved fingers ideal for grasping tree branches. Here was a creature ideally suited to exploit the emerging mosaic of African habitats and to take advantage both of the protection offered by the forests and of the new resources available along the forest edges and in the more open woodlands.

These early hominids were neither apes nor humans, striking instead an unprecedented balance between the old and the new, both in their anatomical structure and in their ecological preferences. This "have your cake and eat it" approach to life proved to be a remarkably stable and successful adaptation, providing the basis for a broad radiation of australopith species, documented throughout sub-Saharan Africa, that only petered out a long while later, well after more modern-looking hominids had appeared. They were evidently sturdy opportunists, able to ride the waves of repeated ecological change.

Still, above the neck the balance between the old and the new was less striking. The australopiths showed (sometimes in exaggerated form) the modifications of the front teeth that were seen in their predecessors and anticipated their successors. But they typically had much larger chewing teeth than are seen either in apes or in humans, and their skull proportions were distinctly apelike, with large faces protruding in front of cranial vaults enclosing brains of modest size. Not for nothing, then, are the australopiths frequently referred to as "bipedal apes." We have a pretty good record of endocasts of these early hominids (indeed, the first australopith to be discovered was a juvenile face attached to a natural brain cast—and it was the form of the brain that initially tipped off its describer, a neuroanatomist, to its owner's hominid status).

Australopith brains are broadly grapefruit-sized, ranging from about 310 cc to perhaps as much as 575 cc. The larger ones generally belonged to members of species believed to have had more bulky bodies, although a shortage of good skeletons makes it hard to demonstrate this for sure. Lucy herself, alas, doesn't boast enough cranial fragments to allow a good estimate of her brain size. But at a best guess it seems unlikely that, as you can see from the figure, any australopith species had an encephalization quotient (EQ, a figure that reflects brain-to-body-size ratio) that differed significantly from that of a bonobo. Indeed, by one calculation the EQs of *Australopithecus afarensis* and of a bonobo come out at an identical 2.44, although later australopith species do tend to be a bit higher (and regular chimpanzees a bit lower). Looking at the brain itself may, on the other hand, tell at least a modestly different story.

The feature of the brain provoking the conclusion that australopiths were human relatives was the position of the lunate sulcus, a fissure at the side of the occipital lobe that marks the front of the visual area. Usually well marked in ape brains, this crease is harder to spot in humans; where it is seen in our species, it lies farther back than in apes. What this indicates is a reduction in surface area of our primary visual cortex relative to the brain as a whole—though of course, given our huge brains, not an absolute reduction. The first australopith endocast was described as having a lunate sulcus positioned to the rear, much like that of humans, and this unleashed an argument that raged for three-quarters of a century, perhaps only partly as the result of undoubted poor definition of sulci in that region of the

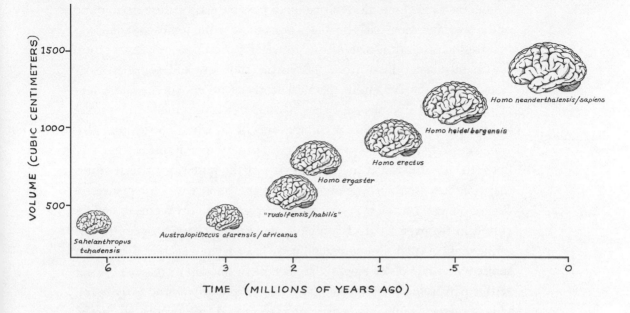

VOLUME (CUBIC CENTIMETERS)

1500 —

1000 —

500 —

Homo neanderthalensis/sapiens

Homo heidelbergensis

Homo erectus

Homo ergaster

"rudolfensis/habilis"

Australopithecus afarensis/africanus

Sahelanthropus tchadensis

TIME (MILLIONS OF YEARS AGO)

6   3   2   1   .5   0

Diagram showing how average hominid brain sizes have increased over the past two million years, after initially flatlining for some five million years. Diagram by Gisselle Garcia.

ancient endocast. Based on new evidence, the original interpretation now appears to have been vindicated, confirming indications from comparative studies that reorganization of the visual system has been a significant theme in human evolution.

The ancient dates of the australopiths suggest, of course, that this trend started early. Neuroscientists have proposed that such modification reflects an enhancement among hominids of the ability to analyze rapidly moving visual stimuli, and this seems reasonable. But the more specific suggestion that it may be related to decoding the lip movements involved in speech seems implausible, if the trend was already under way among the almost certainly nonlinguistic australopiths.

The generally poor definition of cerebral fissuring on their endocasts, together with the traumatic nature of the debate over the lunate sulcus, has largely discouraged detailed speculation on putative brain reorganization among the early hominids. But one other possible early feature of hominid brain reorganization has also been pointed to with some confidence: a relative increase in the size of the posterior parietal cortex. This cortical region integrates inputs from the visual, auditory, and somatosensory (touch) systems to help in the planning and execution of deliberate voluntary movements, and new work has clarified this involvement. In response to external stimuli, it is the "premotor" area of the frontal cortex that generates the "urge" to move on which the primary motor cortex then acts; but it turns out that it is the inferior part of the posterior parietal cortex that produces "sensory representations of the predicted consequences of the movement." In this way, the posterior parietal cortex seems to be "involved in the experience of conscious intention . . . [and] . . . contributes to the sense of controlling our actions." It is an essential component of our subjective feelings of "free will." Exactly how the modest increase in the size of this part of the cortex would have been expressed among early hominids is anybody's guess: speculations have ranged from improved throwing skills through facial recognition to the creation of better mental maps of home ranges. But at the very least, as in the case of the primary visual cortex, the expansion of the posterior parietal cortex hints at the beginnings of functional neuroarchitectural reorganization (and increasing behavioral complexity)—well before the hominid brain began to expand.

Making Stone Tools

One of the big problems we have in interpreting the pattern of evolution of the hominid brain lies in our lack of a firm taxonomic scheme in which to place the observations we can make. By and large, over the past half-century paleoanthropologists have been reluctant to recognize more than a minimum of species. But they now recognize pretty universally that we cannot just arrange the fossil hominid species we know into an orderly progression culminating in our own Homo sapiens. The history of our family turns out, in fact, to have been one of vigorous evolutionary exploration of the hominid potential, whether expressed in the australopith style or in our own. As a result, we know a lot of fossil hominid species, but we can't specify exactly how many there were or what their time ranges might have been. Nor do we know what range of brain sizes any of them encompassed, and (as we've already mentioned) the variation of this feature in the one species we do know (our own) is enormous. All of this makes it a tricky to know exactly what to make of the undisputable fact that, at some time after about two million years ago, hominid brains began to expand, after having essentially flatlined for several million years. The australopiths had survived perfectly well with their ape-size brains and whatever range of behavioral repertoires went with them; and at best the evidence is fairly weak that the latest australopiths had significantly larger brains or more complex behaviors than the earliest ones had had—with one huge exception.

This exception was the manufacture of stone tools, perhaps the most fateful behavioral innovation ever made in hominid history. Without stone tools, hominids were limited essentially by the same factors that circumscribed the existences of all other primate species. With them, everything changed. Just who it was who first thought of banging one piece of rock with another, in exactly the fashion required to detach a sharp cutting flake, we'll never know. Right now the evidence is strongly suggestive but at the same time frustrating. The very first claimed evidence that an animal was butchered using sharp stone flakes comes not from those flakes themselves but from animal bones bearing deep, narrow incisions of the kind known to be made only by such cutting tools. Those cut-marked bones, about 3.4 million years old, come from a place in Ethiopia known as Dikika, just across the Awash River from the site at which Lucy was found. What's more, *Australopithecus afarensis* fossils are also known at Dikika, and this is the only hominid species to which we can reasonably attribute the butchery. Naturally enough, given that the earliest stone tools we actually know are almost a million years younger, the nature of the Dikika cut marks has been disputed—and they may in fact have been made by natural, rather than artificially produced, sharp flakes. But it is surely significant that the next known instance of butchery, also from Ethiopia, equally implicates an australopith toolmaker, this time Lucy's close 2.5-million-year-old relative *Australopithecus garhi*. The evidence in this case is very similar to that from Dikika, but it is less questioned, because it comes from a time from which actual stone tools are known at a handful of sites in eastern Africa. Almost certainly, then, the first stone toolmaker was an australopith.

What is most important here about this new practice is that, as we have seen, stone tool-making is a definite advance over anything any modern ape even has the potential to do. It may seem to us like a simple thing to bash a sharp flake off a piece of stone, but in fact such unimpressive-looking objects are products of a complex behavioral sequence that involves recognizing which rocks are suitable (most aren't), carrying them often long distances to a place where needing them has been anticipated (because you'll most likely be doing your butchery where suitable rock isn't available), and deploying both the cognitive and motor skills necessary to undertake the actual tool manufacture and the subsequent butchery (which is maybe where the posterior parietal cortex comes in). No matter what the other cognitive correlates of these behaviors might be, these by themselves are an enormous leap. Once stone tools were on the scene, the lives of the hominids, and their way of looking at the world around them, had changed forever.

But it's not simply the directly cognitive implications of stone tool-making itself that are relevant here. Also important are the behaviors implied by the use of those tools. It is not known whether the earliest stone toolmakers scavenged or hunted the cadavers of the large animals they butchered; scavenging is an unusual behavior for a primate, but the australopiths were small-bodied creatures that were far from specialized for open-country life, and they hardly fit the image of the mighty hunter. As we've seen, chimpanzees are known to cooperatively hunt small mammals such as colobus monkeys and young bushpig and antelope, and it is entirely plausible that australopiths could and would have done something similar. But even early on, the first stone toolmakers were butchering the carcasses of large animals that it is unlikely they could have run down. Probably the simplest conclusion is that they captured small animal prey and profited from the death by other causes of larger mammals, although the latter activity would have been hazardous in itself, since competition for such resources from professional carnivores would have been stiff. Not a few australopith bones are known with tooth marks bearing witness that these early hominids were not infrequently themselves dinner for leopards and hyenas. Possibly this is where the throwing skills perhaps implicit in their enlarged posterior parietal cortices might have come in, both for protection and in hurling rocks to drive more powerful predators off a kill.

Whatever the exact case, there is growing evidence that australopiths (even ones whose massive grinding teeth would seem on the face of it to indicate a tough plant-based diet) typically incorporated a significant component of animal protein in their food intake. This evidence comes from "you are what you eat" studies of carbon isotope ratios that are preserved in the enamel of australopith teeth. The ratios between lighter and heavier carbon isotopes in living tissues are determined by diet, and the carbon isotopic signal differs most greatly between browsing and grazing animals. There is a remarkably high grazing signal in the enamel of australopith teeth; and since there's no way that australopiths could have got this signal directly by eating grasses, it's much more likely that they had actually been eating small grazing animals and had picked up the signal from them. A potential victim of this kind favored by isotopic scientists is hyraxes, which were abundant in the environments the australopiths favored.

Complexity of lifestyle has, of course, significant implications for cognition all by itself. But perhaps even more important in this case are the energetic implications of meat eating, because the brain is an extremely energy-hungry organ. In *Homo sapiens* the brain accounts for only about 2 percent of our body mass, but it uses up between 20 and 25 percent of all the energy we consume. That is huge! The bigger the brain, of course, the more energy it will consume, and inevitably, any primate species that is going to develop a bigger brain is going to have to find some way of paying for it. Either the brain will have to reduce its energy intake, or its possessor will have to cut down energy use elsewhere. It turns out that our predecessors probably did a bit of both; what is certain is that, without an increase in the quality of their diet, hominids could never have embarked on the trajectory of brain expansion that is such a striking feature of the family following two million years ago. After all, in evolution the necessary conditions have to be in place before innovation can flourish. A prerequisite of bigger brains was a diet of higher quality than the fruit-based food intake typical of chimpanzees living both inside and outside of their preferred forested habitat; so on the basis of the physical evidence alone, it seems pretty probable that a diet enriched by animal fats and proteins preceded both stone tool-making and the upward inflection of the curve of hominid brain enlargement.

Stepping Out

In the half-million years following the appearance of stone tools at about 2.5 million years ago, several fragmentary fossils have been described as "early *Homo*," though their claim to this status is dubious and none of them includes enough pieces of the skull to tell us anything about their brains. You could also group with these bits the 1.8-to-1.6-million-year-old fragments from Tanzania's Olduvai Gorge that have been called *Homo habilis* ("handy man"). Though attributed to *Homo*, these fossils tend to blur the distinction between the australopiths and our own genus. Just after 2 million years ago, however, we do begin to find evidence of new kinds of hominid that justify inclusion in our genus *Homo*. They are basically unanticipated in the fossil record and do not bear witness to anything we could call a "transition" between any known *Australopithecus* and *Homo*. Instead, the appearance of our genus seems more plausibly attributable to a short-term major developmental reorganization due to a shift in gene regulation.

The earliest fossils we can confidently attribute to the genus *Homo* come from Kenya, and the best preserved of these is a pretty complete cranium (the skull minus the jaw) with an internal capacity of some 850 cc. This is a huge increase in size compared to any australopith, and the cranium itself looks much more modern, with a much-reduced face and palate and a better inflated vault. Sadly, we have no associated bones of the skeleton to tell us just how big this individual's brain was in relation to its body size; but in concert with other evidence it tells us that the hominid trend toward large relative brain size was well under way by 1.8 million years ago. Although some traditionalists like to allocate this specimen to the species *Homo erectus*, a species based on a much later fossil from Java that was established back in the nineteenth century as "the" ancient form of *Homo*, this specimen is now more commonly allocated to the species *Homo ergaster*. And by far the best-known member of this species is the "Turkana Boy," a quasi-complete adolescent male skeleton from northern Kenya dating to about 1.6 million years ago. This marvelous specimen gives us an unparalleled insight into hominid life as, for the first time, hominids committed themselves to living in the savanna environments that were opening up and expanding in Africa throughout this period.

The Turkana Boy presents us with a total contrast to the "bipedal apes." No longer hedging his adaptive bets, the boy was tall, slender, and long-limbed, perfectly suited to life away from the shelter of the forests. The climbing adaptations of the australopiths had disappeared, replaced by body proportions that were essentially like our own. Significantly, although the boy had already reached something not far from his adult body size when he died at the developmental stage of a modern fourteen-year-old, he had lived for only eight years, having matured far faster than any modern child. This limited childhood by itself has implications for cognition in *Homo ergaster*, since the extended childhood (and period of helplessness) of *Homo sapiens* allows much more time for learning information essential to economic and social life to be absorbed.

As it was, Turkana Boy's brain size of about 900 cc was close to what it would have been had he survived to adulthood. And with his skeleton readily to hand, we can state with confidence that his EQ showed a distinct increase relative to anything we might estimate for an australopith. Depending on how it's calculated (the official estimate is 4.48), Turkana Boy's EQ is a little under twice that of *Australopithecus afarensis*, although it's still only just above half of yours or ours. The scientists who studied Turkana Boy's endocast were impressed by the presence of an enlarged "Broca's cap," that area on the surface of the left frontal lobe that is traditionally associated with speech production in modern humans. In isolation, this might suggest at least the beginnings of a potential for speech. But the describers of the skeleton noted that the thoracic part of the vertebral canal (the bony tube through which the spinal cord passes to distribute nerves to the body) is rather narrow, suggesting to them that he had possessed fewer nerves controlling the chest area. And this in turn implied that he had lacked the fine control of breathing on which we depend to produce to sounds of speech. Still, the betting nowadays is that Broca's area is generally involved with the fine motor control of the right hand and of the components of the vocal apparatus, rather than specifically with speech itself. And if this is the case, it is plausible that the enlargement of this cortical area that expressed in Turkana Boy's endocast might actually have been associated principally with dexterity and fine motor control of the right hand. The probability that this was the case is enhanced by the marked asymmetries between the two hemispheres of the brain, of the kind known to be associated with right-handedness in humans.

Perhaps this is hardly astonishing, since experimental archaeologists have shown that from the very beginning stone tools were more commonly made by right-handers than by lefties. What is more surprising, perhaps, is that with the appearance of the new and improved hominid form exemplified by Turkana Boy, no new types of tool appeared. Because, for all their larger brains and other physical innovations, Turkana Boy and his kind continued to make crude cutting and pounding tools that were more or less identical to those their more archaically proportioned predecessors had been making for a million years or more. It was not until about 1.5 million years ago that a new concept in stone tool-making was generally adopted, though a recent report puts its first appearance as early as 1.78 million years ago. This innovation was the "hand ax," a much larger implement (typically about 20 centimeters or more in length) that was carefully fashioned on both sides to a standard symmetrical teardrop shape. Such tools were produced in enormous quantities, sometimes at "workshops" next to sources of particularly good stone.

In the hand ax itself, and in its processes of manufacture, we can see an enormous conceptual leap. The early stone toolmakers had merely been after an attribute—namely, that sharp cutting edge. It clearly hadn't mattered to them what the resulting flakes actually looked like. With the hand ax, in contrast, we see a desire to impose form on the external world and the employment of a complex series of procedures to realize a shape that had existed in the maker's mind before knapping started. Furthermore, the existence of the workshop sites strongly implies a degree of economic specialization and the differentiation of social roles. All this is dimly glimpsed, unfortunately, because we have little material evidence to testify to how

A hand ax (left) and a cleaver from the Acheulean site of Saint-Acheul, France. Photo by Willard Whitson.

the hand-ax-makers' lives may have changed and become more complex in other ways.

The new way of viewing the world embodied by the hand ax was clearly made possible by the enlarged brain of *Homo ergaster* and by the increasing motor skills that went along with it. But clearly, given the long delay in the invention of the new technology, it was not mandated by these new acquisitions. Like any new technology, the hand ax had to be invented by hominids that already possessed the cognitive wherewithal to conceive of and execute it; in other words, the potential for this new behavior had to be in place *before* it was realized, in an *exaptive* rather than an *adaptive* context, as something merely *available* to be coopted for a new use—or not.

This idea brings us back to the role of chance in the evolutionary process, and it helps us understand a durable pattern in human cognitive evolution: you can never confidently explain the origin of a new behavior or technology by the arrival of a new kind of hominid. This had, of course, been true from the beginning, given that the first stone tool-makers were bipedal apes of a general kind that had already been around for a long time. And it underwrites the important fact that the evolution of human cognition has not been a process of gradual fine-tuning over the eons. Instead, it has involved a haphazard set of biological acquisitions that were made sporadically over several millions of years, each of them discovered as a behavioral advantage only in retrospect. This does a lot to explain the "messy" aspect of our modern human brains, about which we wrote at some length earlier.

Taking Over the Old World

Well before the hand ax appeared on the scene—indeed, only fractionally later than the appearance of *Homo ergaster* itself—another momentous event had occurred. This was the first spread of hominids beyond the confines of their parent continent, Africa. The first good evidence we have of this exodus comes from the unexpected locality of Dmanisi, in the Republic of Georgia between the Black and Caspian Seas. Freshwater sediments at Dmanisi, which were deposited on a lava flow laid down around 1.8 million years ago, have yielded an extraordinary set of hominid fossils from this time. They have also produced abundant early artifacts, all of them primitive cutting flakes and the cores they from which they were knocked. Perhaps it's not so surprising, then, that the four hominid crania so far described from Dmanisi boast very small brains, ranging in size from about 600 cc to about 775 cc. Their endocasts have yet to be described, but a pair of partial skeletons suggests that these hominids were substantially shorter in stature than Turkana Boy, though the bones themselves are said to suggest a relatively modern body structure. It is thus possible to suggest that the spread of hominids out of Africa was made possible by the arrival of the essentially modern body form and facilitated by an expansion of more open habitats in western Asia similar to that occurring at the same time in Africa. The Dmanisi crania look oddly primitive, and they plausibly represent the descendants of very early modern-bodied African émigrés of the genus *Homo*—although they are usually allocated to *Homo erectus* in its broadest sense, which is no more than a useful way of avoiding a potential taxonomic problem.

On the behavioral front, an intriguing observation is that one of the Dmanisi hominids, an old male, had lost almost all of his teeth well before he died. When an individual loses a tooth, the bone of the jaw that had accommodated its roots is also eventually lost, in a process that can take a long time. Because this individual had very little root-supporting bone left, we know he had been virtually toothless for years before his demise. This has led to speculation that he had required the assistance of his fellows in processing the food he took in; though there is no way to prove this, his lack of teeth is strongly suggestive of social support. This in turn implies powerful feelings of empathy, and possibly also of some considerable complexity of social organization, among the Dmanisi hominids.

By not long after Dmanisi times, hominids had spread all the way to eastern Asia, where the familiar *Homo erectus* is known at least as far back as 1.7 to 1.6 million years ago. What we know of them and their brains doesn't add much to what we know of *Homo ergaster* in Africa, where sporadic finds over the period between about 1.5 and 1 million years ago seem to suggest that hominid brain sizes were continuing to increase modestly over the period. Just what this means is not entirely clear, partly because of the deficiencies of our current taxonomic scheme. As we have seen, if hominid brain sizes are plotted against time on a chart, the curve remains pretty flat until about 2 million years ago, when its slope increases dramatically. Before 2 million years ago, brains were basically ape-sized; a million years ago they had doubled in size; and by about 200,000 years ago they had doubled again. It is tempting just to join up the dots to form a straight line suggesting an inexorable increase in the hominid lineage from 2 million years ago on. Yet we also know that there is a lot of anatomical variety among hominids of this period, and it is almost certainly highly misleading to lump all this diversity into the single species *Homo erectus*, as many paleoanthropologists like to do.

If, like us, one prefers to detect a diversity of hominid species in this period, the picture changes enormously. As far as we can tell from a rather spotty sample of hominid crania, there is a pretty good range of brain sizes out there at any one time. Viewing this range through the lens of species diversity makes it much more compelling to see the average brain size increase among members of the hominid family during this time as due to the preferential survival of larger-brained species, rather than to inexorable brain enlargement in one central lineage.

Until the taxonomic picture is solved to general satisfaction, we really won't know what the actual pattern of hominid brain size increase over the past two million years was. But it is almost certainly significant that we *can* demonstrate that brains got larger in at least three independent lineages (the long-lived *Homo erectus* group in Asia, and in the lineages leading to the Neanderthals in Europe and to *Homo sapiens* in Africa: we'll say more about the latter two groups in a moment). That brain size increase occurred independently in this many lineages, and maybe more, argues strongly that there is something about belonging to the genus *Homo* that has strongly predisposed all its members toward increased brain size. That something may lie in the general lifestyle that originated about two million years ago on the expanding African savannas, where hominids were almost certainly both hunters and prey; but, whatever the common element was, we shall have to identify it before we can claim to have any full understanding of our own neural and cognitive evolution.

The First Cosmopolitan Hominid

The evolutionary—if not the cognitive—picture starts to get somewhat clearer at just over six hundred thousand years ago, when we begin to encounter fossils of the earliest incontestably cosmopolitan hominid species, *Homo heidelbergensis*. Named for its first-discovered member, now also known to be its oldest, this species is represented at sites in eastern and southern Africa, in Europe, and in China, thus covering most of the habitable Old World. Most of these sites are poorly dated, but some may be as young as about two hundred thousand years ago. In terms of skull structure *Homo heidelbergensis* anticipated the future to a much greater degree than *Homo ergaster* had, and although its facial skeleton was quite massive, it had a significantly larger brain than any of its predecessors. Measurable endocasts come in at between about 1,150 cc and about 1,325 cc; all are well within the range of *Homo sapiens*, and the largest is not far below the modern average. EQs are thus up close to modern levels, the mean value for the species being over 80 percent of the modern average.

Paleoneurologists have inclined toward being impressed by the similarities of *Homo heidelbergensis* endocasts to modern ones, rather than by the differences, perhaps because of uncertainty over what those differences might mean—if indeed they are significant at all. All endocasts of this species so far described are said to show expanded Broca's areas; otherwise the literature is disappointingly mute, beyond recording typical cerebral asymmetries and a general broadness and flatness of the frontal lobes. This is where our lack of knowledge about internal organization of the various cortical areas visible in the endocasts becomes enormously frustrating. The larger our fossil relatives' brains get, and the closer in appearance they become to our own, the harder it becomes to speculate about potential cognitive differences on the basis of external shape. Thus, in the case of *Homo heidelbergensis*, nobody seems to have any idea what such features as the exact form of the frontal lobes may be telling us. Did these hominids have what we have, only less of it on average? Were their brains somehow qualitatively different from ours in their internal structure or in the way they functioned? After all, they were housed in substantially different-looking braincases. Or are we not looking at any meaningful structural differences at all? Questions like these will not be answered by looking at endocasts, at least at this stage of our ability to interpret them. This means that the only route we have toward developing a picture of what *Homo heidelbergensis* and its later extinct relatives were actually like, as cognitively functioning beings, is by looking at the material record of their behaviors that they left behind. Fortunately, that record is quite suggestive.

The archaeological records associated with the earliest known *Homo heidelbergensis* fossils are disappointingly sparse, and they are generally unimpressive where they exist at all. It is, however, within the tenure of *Homo heidelbergensis* that we begin to pick up evidence of significant additions to the hominid behavioral repertoire. Among these are the first traces we have of deliberately constructed dwellings, sadly (for us) without any fossil remains of the occupants. At a place called Terra Amata on the Mediterranean coast of France, the remains were excavated of a seasonally occupied beach camp dating from almost four hundred thousand years ago. Features of the site included what appear to have been the foundations of several large huts, as indicated by large oval rings of stones used to reinforce rows of cut saplings that had been stuck into the ground. It is believed that the saplings had been gathered at the top to form an enclosed space and that breaks in the stone rows indicated entrances to the huts. Within the entrances fires had burned in scooped-out hearths, as indicated by blackened cobbles and animal bones.

This evidence of the domestication of fire, and of its use in cooking, was until recently one of the earliest such indications on record. But in 2004 an almost eight-hundred-thousand-year-old site was reported in Israel that showed clear signs of repeated hearth use, as indicated by thick layers of carbon in the deposits. This is curious, because the domestication of fire must have been a hugely formative event in hominid life. So it is odd that such an innovation should not have caught on rapidly. Yet there is as yet no evidence for fires burning in hearths anywhere between that early occurrence and Terra Amata times, following which evidence for regular fire use begins to pick up. Still, the archaeological record is oddly selective, and there are indeed scientists who believe that the commitment of early *Homo* to the savanna can have been made possible only by the control of fire—with all the cognitive implications that anyone can readily imagine. In the absence of undisputable indications of hominid fire use before that Israeli occurrence, the evidence for this proposition is inevitably circumstantial, but it certainly has its compelling aspects.

For one thing, the open savanna is a dangerous place for a slow and relatively defenseless hominid to be—especially at night, when fire would have been most valuable in discouraging the huge variety of predators that abounded in this habitat. For another, since far fewer plant resources would have been available to early hominids on the savanna than in the woodlands, these creatures would necessarily have been substantially dependent on animal proteins: the more so because of the energetic demands made by their enlarging brains. It has been argued that the need for additional energy might have been mitigated at least to some extent by a compensatory reduction in equally metabolically expensive digestive tissues; certainly, the anatomical indications are that the australopiths had the large guts of fruiteaters, while early *Homo* had reduced intestines, as we do today. This physical reduction could have been made to work only by increasing the animal component of the diet—and in the absence of a digestive tract specialized for carnivory, the best way of making raw meat digestible would have been to cook it. All in all, an attractive argument; but it remains the case that we don't begin to find actual evidence for the routine use of fire until Terra Amata times and beyond.

That four-hundred-thousand-year point is also when we find the first evidence for the making of wooden spears, at the site of Schöningen, in northern Germany. This is, of course, a tricky area: wood preserves only in the most exceptional of circumstances, and we have evidence from characteristic wear on stone tools that hominids had been cutting wood much earlier. And, as we'll see, *Homo heidelbergensis* is not the only possible candidate for making such tools at this juncture. Still, the Schöningen spears are exceptional, all the more so for being found in association with the butchered remains of horses. Several of these remarkable implements were preserved in a peat bog, most of them well over two meters long. What's more, they are carefully shaped to the form of javelins, with the weight concentrated at the front, so they resemble throwing spears rather than thrusting implements. If indeed they *were* thrown, they imply that their makers indulged in the cognitively complex activity of ambush hunting, a sophisticated activity that many thought hominids had not invented until much later in time. Of course there are, as always, complications: the spears are shaped to a point, not tipped with stone; and it's been argued that if hurled from any distance they would have uselessly bounced off any thick-skinned mammal. Still, perhaps equally suggestive are some more modest pieces of wood that seem to have been grooved as handles for flint flakes—making them the earliest composite tools known, and evidence for yet another conceptual advancement.

It is also within the tenure of *Homo heidelbergensis* that we find the next major innovation in stone working. This was the invention of "prepared-core" tools, which show up in Europe at around three hundred thousand years ago, although variants are occasionally found even earlier in Africa. Tools of this kind are made by carefully shaping a nucleus of stone until a final blow detaches a more or less finished tool of predetermined shape that has the advantage of a continuous cutting edge around it. Again, we are clearly witnessing a whole new level of complexity in conceptualizing what can be done with a piece of stone, and this advancement sits quite easily with the emerging picture of hominid life during this period. This was the time of the Ice Ages, during which periods of extreme cold afflicted the northern continents at intervals of around a hundred thousand years, causing the advance and subsequent retreat of the polar ice caps. European and Asian hominids thus had to cope with intermittent (and often quite rapidly oscillating) periods of extremely cold climate, while their tropical African relatives experienced droughts and periods of equal climatic instability. As a species with an Old World–wide distribution, *Homo heidelbergensis* as a whole faced an unusually wide range of environmental conditions; cultural accommodation to extremes must certainly have been part of its members' armamentarium for coping with it.

Yet over the entire time and space it occupied, *Homo heidelbergensis* left behind no artifacts or behavioral traces that we can convincingly interpret as symbolic. As sophisticated as its members undoubtedly were in exploiting and dealing with the world around them, there is no indication that they processed information about it in the way we do today. This is not to demean them: intuitive, nonsymbolic mental processes can produce complex and subtle responses to stimuli coming in from the environment, and there can be no doubt that *Homo heidelbergensis* was capable of responses of this kind. The cognitive skills of members of this species were surely as impressive as their motor ones. But whatever paleoneurologists may make of their endocasts, there is no evidence whatever that these hominids remade the world in their heads. That peculiar ability was still to materialize.

Neanderthals and Cro-Magnons

Even more accomplished than *Homo heidelbergensis* was *Homo neanderthalensis*, a distinctive hominid species that lived in Europe and western Asia between about two hundred thousand and thirty thousand years ago. The offshoot of an endemic European lineage that had roots going back to over six hundred thousand years ago (and that apparently had coexisted for a long while in Europe with *Homo heidelbergensis*, which it seems eventually to have outcompeted), *Homo neanderthalensis* boasted a brain as large as our own. Indeed, at 1,740 cc, the largest-brained hominid fossil we know is that of a Neanderthal and not of an early *Homo sapiens*. Their large neural masses gave the Neanderthals an EQ that has been calculated at over 99 percent of its modern human equivalent, given that they were on average of slightly shorter stature but of more robust body build than we are today. The

Neanderthals' big brains were, however, housed in a distinctively different skull that was perched atop a body of equally distinctive proportions, testimony to a long period of independent evolution since their lineage and ours shared a (probably African) common ancestor.

In contrast to the high, vaulted skull of modern *Homo sapiens*, flexed to tuck a small face beneath its front, the adult Neanderthal braincase is long, broad, and low, and the brain inside it is shaped to match. Once again, the paleoneurologists have not observed anything in the endocasts to suggest any major functional differences, but it has now been documented that the adult brain shape discrepancies in the two species result from very different developmental trajectories. Like other extinct hominids, Neanderthals seem in general to have matured faster than modern children do, limiting the time available to them to develop social and economic skills. But in the case of the brain specifically, modern humans rapidly do something different in early postnatal development.

It seems that the brains of both Neanderthal and modern newborns shared pretty much the same shape, helping both species with their common problem of getting a large-headed offspring through the birth canal. In their first year of postnatal life, however, infant *Homo sapiens* depart abruptly from the standard pattern of development that was evidently retained by the Neanderthals. At this early stage human babies undergo a very fast "globularization" in brain shape, as the distinctive modern cranial proportions are attained. Exactly what this ontogenetic difference might have meant in terms of cognitive development in the two species, or in their adult brain organization, is not clear. But this distinctive developmental discrepancy does help emphasize that *Homo sapiens* has very recently acquired a radically new adult anatomy, one that distinguishes it strongly from all other primates, including its closest extinct relatives.

Because we find it very hard to interpret the undoubted differences in proportion between the equally large Neanderthal and modern cranial endocasts, we have to abandon paleoneurology and turn to other kinds of evidence if we wish to make inferences about how the possessors of these brains viewed and responded to the world. Fortunately, the Neanderthals have left us a copious record of their activities; perhaps even

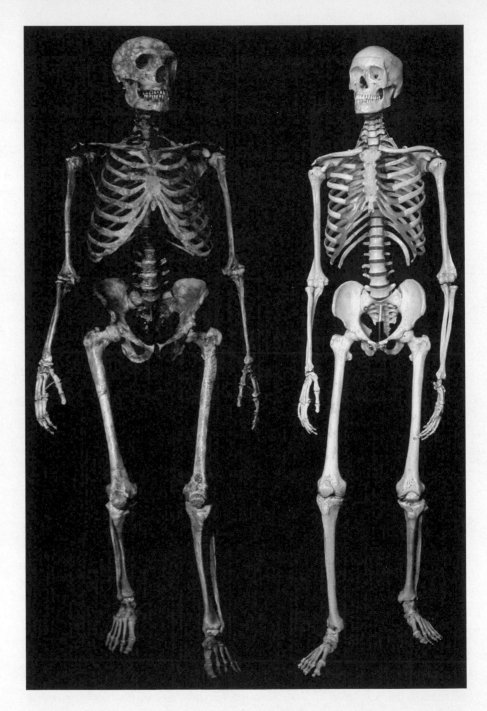

Comparison of a reconstructed Neanderthal skeleton (left) with that of a modern human of similar stature. Apart from the cranial differences, note especially the very different structures of the thoracic and pelvic regions. Photo by Ken Mowbray.

more fortunately, the fact that *Homo neanderthalensis* and *Homo sapiens* at least transiently shared the same region of the world gives us some insight into the different ways in which the two species coped with similar environmental circumstances. Importantly, *Homo neanderthalensis* was cognitively very complex, possibly in its early days the most accomplished hominid species there had ever been. Neanderthals were extremely competent craftsmen and perhaps the most skilled practitioners ever of the prepared-core method. In terms of motor skills, they had it all.

They were also very flexible and resourceful exploiters of their environment or, rather, environments, for they flourished throughout a vast area of Europe and western Asia over a long period during which climatic conditions fluctuated enormously. At least at certain times and places Neanderthals were specialized hunters of some of the most fearsome animals around, woolly mammoth and woolly rhinoceros among them; at others they eked out a living from much humbler resources. Neanderthals at least occasionally buried their dead; they rigged up shelters against the elements; using fire was a part of their daily routine; and there is no doubt that they produced some very significant cultural responses to some extremely difficult environmental conditions. These must have included the invention of clothing, for it has been calculated that without it and other cultural accoutrements they could never have survived the severest cold stresses of the last Ice Age. To have accomplished the trick by the physical means of adding subcutaneous body fat would have involved an impossible near-doubling of the Neanderthals' body weight.

Yet impressive as they are, these cultural compensations are all things that could have been achieved by the exercise of a powerful intuitive intelligence. And for all of their undoubted attainments, Neanderthals did not anywhere show the creative spark that so clearly animated the lives of the "Cro-Magnons," the modern *Homo sapiens* who began to trickle in to the Neanderthals' European heartland beginning around forty thousand years ago. Not much more than ten thousand years later, the Neanderthals were gone, and although it has been argued that Neanderthal populations were already declining before the Cro-Magnons arrived, it remains unlikely that this was pure coincidence. Much more plausibly, the replacement was due at least in part to the unusual qualities of the newcomers. The Neanderthals simply did what their forerunners had done, if perhaps a little better; they evidently had not taken the qualitative leap to symbolic ways of processing information that so thoroughly transformed the Cro-Magnons' relationship with the world around them.

Of course, symbolic cognition is something that we cannot directly observe in any extinct species: it is something that has to be inferred from proxy evidence that is preserved in the tangible record, and this is undoubtedly a limiting factor in what we will ever be able to say. But although we will see in a moment that it is perhaps going too far to claim that no Ice Age technology, however adroitly carried out, can be taken on its own as proxy evidence of symbolic thought processes, it remains true that technology of the kind possessed by the Neanderthals can hardly be taken as such evidence. For example, when they are carefully made, the Neanderthals' "Mousterian" tools are things of such beauty that some people have had difficulty believing that the technology needed to produce them could ever have been passed down through the generations in the absence of language. But when some Japanese researchers tried teaching naive subjects (of course, symbolic *Homo sapiens*) to make such objects both with and without the benefit of linguistic explanation, they found that the addition of language made no difference.

The opposite result would have been telling, for language is without any doubt the ultimate symbolic activity, involving as it does the abstract representation of the environment as a vocabulary of discrete symbols, whose recombination in the mind stands as the fount of the human creative process. It is the ability to make new mental associations in this way that makes us the extraordinary creatures that we are. And the Neanderthals left behind very little, if anything, that may unequivocally be interpreted as a symbolic object. Some very late objects from a site in France that do appear to be legitimately interpreted as symbolic, and were widely viewed as the work, or at least the possessions, of Neanderthals, have recently been shown to be very dubiously associated with *Homo neanderthalensis*. Their subtraction from the Neanderthal oeuvre has left us with precious little in the way of Neanderthal symbolism to point to.

What's more, even in arguable cases one swallow doesn't make a summer; and there can be no doubt whatever that symbolism was never a routine feature of Neanderthal life, anywhere in the entire vast tract of time and geography these hominids inhabited. All this contrasts dramatically with the lives of the Cro-Magnons, which were literally drenched in symbol. These were the people, modern in every respect, who about thirty-five thousand years ago (several thousand years before the extinction of the Neanderthals) began to create the great paintings of such renowned Ice Age caves as Chauvet, Lascaux, and Altamira—art as powerful and sophisticated as anything ever achieved since. At the same time, the Cro-Magnons produced musical instruments, elegant and sometimes naughty small carvings, delicate and beautifully observed animal engravings, and systems of notation recorded both on cave walls and on tiny bone plaques. Perhaps equally telling of a new sensibility is a sudden acceleration in the pace of innovation. Previously, vast spans of time had elapsed between major material innovations; now, one technology succeeded another at an ever-increasing pace. What this seems to reflect is an entirely new approach to dealing with life's problems. Hominids had previously tended to use old tools in new ways as environmental circumstances changed, but *Homo sapiens* was now reacting to new exigencies by inventing entirely novel technologies. This change had extensive economic ramifications: whereas Neanderthals had been pretty thin on the landscape, the Cro-Magnons populated it in far denser numbers, indicating pretty clearly that they were exploiting the landscape around them in vastly more intensive ways.

All of these Cro-Magnon material innovations are compelling evidence for the arrival of a completely new kind of sensibility: the emergence of a mind capable of processing stimuli from the environment in an unprecedented fashion. In short, the huge contrast between the Neanderthals and the Cro-Magnons does not simply reflect the kind of difference that you might expect to find between two average hominid species. Instead, we see a contrast between brains that, although identical in size, *worked* in fundamentally different ways. The Cro-Magnons were not simply an improvement on the more primitive hominid pattern but embodied an entirely unprecedented cognitive model. Whether this difference may have had

Monochrome rendering of a now badly faded polychrome wall painting, probably about twenty thousand years old, in the cave of Font-de-Gaume, France. A female reindeer kneels before a male that is leaning forward and delicately licking her forehead. Drawing by Diana Salles after a rendering by Henri Breuil.

anything to do with the minor distinctions in their external brain proportions that are typically dismissed by the paleoneurologists is something we will look at in a moment.

The Emergence of Modern Behavior

The new sensibility that we see reflected in the great Ice Age cave art of France and Spain was not developed in those places. Instead, the arriving Cro-Magnons brought it with them, having acquired it somewhere else entirely. This other place was almost certainly ultimately in Africa, the continent that had from the beginning been the major source of innovation in hominid evolution.

The earliest fossils we have that bear the stamp of the unique modern morphology come from the sites of Omo Kibish and Herto in Ethiopia. A fragmentary skull from Omo Kibish is believed to date to about 195,000 years ago, and an adult and a child cranium from Herto are quite firmly dated to around 160,000 years. Both were basically modern-looking forms, with tiny faces tucked below globular, vaulted *Homo sapiens*–like braincases. And although their endocasts remain undescribed, the crania certainly contained brains of entirely modern external proportions. The adult from Herto has a published cranial volume of 1,450 cc, comfortably above the modern average. Together these fossils provide evidence for the arrival of hominids with fully modern anatomy at a pretty remote period in time. Yet for all their anatomical innovativeness, their archaeological associations are strikingly unimpressive: "nondescript" stone flakes in the case of the Omo specimen and some of the latest known African hand axes at Herto. Whatever the appearance or potential of those individuals' brains, they certainly do not appear to have been working in a detectably new way.

The same is also true for the first known occurrence of the anatomically distinctive *Homo sapiens* outside Africa. This is in Israel, where numerous sites testify to the presence, at least intermittently, of both *Homo neanderthalensis* and *Homo sapiens* during the one hundred thousand years preceding the last record there of a Neanderthal, at a little over forty thousand years ago. At the site of Jebel Qafzeh the remains of a roughly hundred-thousand-year-old anatomically modern individual lay in clear association with a Mousterian tool kit that was effectively identical to what was being produced by Neanderthals in the same region. What's more, intense scrutiny of the Mousterian produced by the two species in Israel has not turned up firm evidence of any functional difference between the two tool kits. Apparently, over the entire long period during which they contrived to share the Levantine landscape, Neanderthals and moderns were behaving in much the same fashion. This is a striking contrast to what we see in Europe, where behaviorally distinctive modern humans replaced the resident Neanderthals over only a few millennia; it suggests quite strongly that *Homo sapiens* achieved its "edge" only with the acquisition of behavioral modernity.

Partly because of this, most authorities now reckon that the early Israeli *Homo sapiens* represents a "failed" early foray out of the parent continent and that permanent occupancy outside Africa was achieved only much later, by emerging cognitively modern humans. This fits well with what we are learning from a vigorously expanding archaeological record. Although occasional "advanced" tools, such as long, thin stone blades comparable to those produced by the Cro-Magnons, are found at African sites several hundred thousand years old, signs of a significantly new approach to life do not show up until the technological period known as the Middle Stone Age was well under way. The Middle Stone Age was long thought of as the African equivalent of the contemporaneous Middle Paleolithic of the Neanderthals in Europe. However, although there are common elements—notably the use of prepared-core techniques—it is clear that a lot more was astir in Africa than was going on in Europe at the time. This is particularly true of the period following about a hundred thousand years ago.

The best-known Middle Stone Age sites are two caves situated on the southern coast of Africa: Blombos and Pinnacle Point. Blombos is most famous as the place which has yielded the world's first overtly symbolic objects (a couple of smoothed ocher plaques engraved with geometric designs) in deposits about seventy-seven thousand years old. The grinding of ocher is an older practice than this—and Neanderthals did it, too—but the planned symmetrical design is the clear product of a symbolic thought process. This interpretation is reinforced by the finding in Blombos deposits of about the same age of carefully crafted bone tools, later to be a Cro-Magnon specialty, and of numerous shells of small marine snails that were pierced for stringing and bear traces of ocher. If these shells were indeed strung into a necklace or some other form of bodily adornment, they are highly significant, because in all modern human societies body decorations are quintessentially symbolic, typically loaded with overtones of status, occupation, and class. That these "beads" were no exception for the period is confirmed by similar, and even older, findings at sites as far afield as Morocco and even the Levant. On the technological front, it was reported in 2010 that some spear points from Blombos seem to have been "pressure-flaked," using a subtle stone-working technique that even the Cro-Magnons did not practice until about twenty thousand years ago.

Even more dramatic technological evidence comes from deposits more than seventy thousand years old at Pinnacle Point. These have produced tools made from fire-hardened silcrete, a silica-rich by-product of soil formation that is normally a poor material for tool-making. But its qualities improve miraculously when it is subjected to an elaborate multistage process of heating and cooling: one that requires the kind of planning and conceptualization that only a symbolic mind could envisage. If any Old Stone

Age technology can serve as a proxy for symbolic thought, this is it. People may even have been heat-treating silcrete at Pinnacle Point as back far as 164,000 years ago—still in the range of anatomical *Homo sapiens*—but the evidence for this is inconclusive. Yet at that early stage those ancient inhabitants of the cave at Pinnacle Point were definitely already eating marine resources, including a wide variety of shellfish. These had previously been rare or absent in hominid diets, and their exploitation is thought to indicate a new level of sophistication in the hunting and gathering tradition. It is unfortunate that at Pinnacle Point we have no fossils to testify to the identity of the hominids involved and that at Blombos the few hominid teeth known are too badly broken to say much. Still, there is nothing about them to make one think that they are not perfectly reasonably attributable to *Homo sapiens*.

Putting all this together with other evidence, it seems quite reasonable to propose that, following the appearance in Africa of *Homo sapiens* as an anatomical entity, many thousands of years elapsed before the new species began, haltingly, to "discover" its new cognitive potential. This was clearly not a one-time event. Not long after the Blombos plaques were engraved, the entire southern region of Africa was plunged into drought and was apparently largely or entirely depopulated of humans; it may thus seem a bit unlikely that this was the center from which an irresistible wave-front of newly symbolic hominids emerged to take over the world. But whatever the exact case, these South African manifestations are emblematic of the fact that something of unprecedented cognitive importance was astir among hominids of the Middle Stone Age, during which we see evidence for the development of other complex behaviors such as flint-mining, the long-distance trading or exchanging of valued materials, and object decoration.

It seems reasonable to conclude that the emergence of complex symbolic cognition was permitted by some kind of internal rewiring of the brain: a change that was most plausibly acquired in the evidently major developmental reorganization that took place at the origin of anatomically recognizable *Homo sapiens*. What these indicators seem to suggest is that, in the period following (and perhaps even before) about a hundred thousand years ago, the newly arrived hominid species was beginning to experiment with its new cognitive potential—a process that, indeed, continues today. Since the biological underpinnings of the new ability were necessarily already in place, the expression of that potential must have been caused by a cultural stimulus of some kind.

Various candidate stimuli have been proposed, most revolving around the growing complexity of hominids' social lives: increasing levels of intentionality (that ability to read others' minds) are a favorite. But stimuli based on social pressures of this sort would imply a gradual process of acquisition, and the indications we have are that the acquisition of symbolic consciousness in *Homo sapiens* was more or less instantaneous in evolutionary terms. A much better candidate would, then, be the invention of language, which as we've said is the ultimate symbolic mental activity, which we know can be spontaneously generated, and which would necessarily have been a short-term event. Once one human population had attained language, one can readily see how its use would rapidly have spread among adjacent populations that, after all, already had the biological potential to acquire it. And placing language in this context has the advantage that it requires no explanation of how the peripheral apparatus enabling speech, its expression, had been acquired. This is because the requisite anatomy had clearly been in place since the origin of *Homo sapiens*, having obviously been acquired initially in an entirely exaptive context.

A scenario of this kind fits well with what molecular biologists have concluded about the pattern of human spread across the world, based on genomic comparisons among modern populations. From patterns of variation in the mitochondrial and Y-chromosome genomes, it is reckoned that the human species underwent a dramatic contraction—in genetic jargon, a "bottlenecking," about sixty to eighty thousand years ago. Quite possibly because of battering by environmental circumstances in the rigorously fluctuating environments of Africa during this period, the entire species *Homo sapiens* was reduced to a population of between several hundred and a few thousand individuals. If language and symbolic thought had been achieved within this remnant population, as a random and emergent event rather than as an adaptive one, it would explain the subsequent explosive expansion of our species out of Africa, across the Old World, and eventually into the New World. The initial expansion might have been permitted by an amelioration of climatic conditions, but perhaps more plausibly it was due to the new kinds of reaction to circumstances that were enabled by the radically new cognitive style. Still, the fact that earlier tentative explorations of the new capacity may well have ended in failure, as climatic fluctuations periodically rendered large tracts of Africa entirely uninhabitable, suggests that external conditions played an important role.

The ultimate cognitive leap (the transition from a nonsymbolic, non-linguistic state to a symbolic, linguistic one) is an absolutely mind-boggling switch of mental conditions. And indeed, the only reason we have for thinking that such a switch *could* ever have been achieved is that, quite obviously, it *was* achieved. It remains conceivable that the change was enabled simply by the final increase in brain mass to modern proportions. But, especially given the vast range in brain sizes among cognitively normal modern humans, it is much more likely that internal brain modifications were involved. What might these have been?

How Did It Happen?

Much as technological innovations like real-time brain imaging have taught us about what is going on in our heads when we do various cognitive tasks, we still have no idea at all about how a mass of electrochemical signals in the brain gives rise to what we individually experience as our consciousness. And without knowing this, we are reduced to sheer speculation when it comes to knowing exactly what physical ingredient it was that, acquired but not yet exploited at the origin of our species, permits our unique style of cognition. As we have seen, the probabilities are against the notion that this new ingredient is a quantitative one, simply a passive result of increasing brain mass. The Neanderthals had brains as large as ours, but theirs evidently dealt with incoming stimuli in a very different way. It remains possible that a young Neanderthal, confronted with the appropriate cultural milieu, would have been able to achieve a modern cognitive condition; if, however, significantly new neural pathways were involved, this is unlikely. More probably, Neanderthals possessed a purely intuitive intelligence, albeit in a powerful form. Lacking symbolic cognition, they would also have lacked language as it is familiar to us, although undoubtedly they used so-phisticated vocal communication, supplemented with the kinds of gestural and body-language communication that we also employ—though we are rarely consciously aware of using them.

The great neurologist Norman Geschwind proposed in the 1960s that it was the naming of objects (their discrete mental identification) that enabled language—and, through the association we make here, symbolic cognition. Geschwind thought that this association was made possible by an anatomical ability to make direct associations between representations formed in different areas of the cortex directly, without the intermediation of the limbic system and the resulting addition of emotional responses. He identified the brain structure involved as the angular gyrus, a section of cor-tex that lies in the parietal lobe, next to its junction with the temporal and occipital areas. The angular gyrus is large in humans, whereas it is small or absent in other primates, and research has shown that in humans it is im-portant in the comprehension of metaphors. This, of course, encapsulates what language involves: making connections and, what's more, making them according to rules rather than emotions. Unfortunately, the leading paleoneu-rologist Ralph Holloway reports that this region is hard to detect on endo-casts, because the sulci that define it in the living brain almost never imprint on the inside of the braincase. As he neatly puts it, although "you can 'see' it" on endocasts, you cannot "delineate it." So there is no good way to detect whether this highly important region was expanded in any fossil hominid.

The linguist Philip Lieberman thinks that fully modern language was a very recent acquisition in the human lineage, perhaps between ninety and fifty thousand years ago. He argues that fine control of the structures per-mitting speech must have been achieved by coopting preexisting motor structures in the brain that included areas not only of the prefrontal cortex but of the underlying basal ganglia and cerebellum. This makes excellent sense in terms of evolutionary mechanism. The neuropsychologist Freder-ick Coolidge and the anthropologist Thomas Wynn view the full achieve-ment of modern cognition as perhaps even more recent. Harking back to the classic case of Phineas Gage, who as you will recall lost the ability to control his emotions when that iron rod was blown through his frontal lobes, they trace humankind's extraordinary cognitive capacities to improvements in the "executive" functions of the prefrontal cortex that govern decision-making, goal formation, planning, and so forth. With its extensive connections to other brain regions, the prefrontal portion of the cortex, more than any other part of the brain, coordinates what the entire brain is doing at any particular time. Coolidge and Wynn see as critical to higher executive function an increase in "working memory": that set of systems, resident primarily in the frontal lobes, which govern the capacity to hold and integrate sequences of stimuli coming in at a particular point in time.

Based on an analysis of the archaeological record, Coolidge and Wynn discern two major cognitive leaps in human evolution. The first occurred when *Homo ergaster* and its like embarked on an entirely new ecological adventure that required a totally new approach to the exigencies of life. The other happened when "completely modern thinking" was acquired, enabled by a dramatic expansion in working memory and most dramatically expressed in the phenomenal outpourings of Cro-Magnon art. They have even argued that, by the strictest standards of recognition, fully modern cognition can only be recognized universally much more recently yet. But once again Coolidge and Wynn, like Lieberman, are dealing with proxy evidence; and the changes that according to all three authors underpin cognitive modernity would not necessarily be expected to show up in the kind of physical features reflected in endocasts. What's more, even though enhanced working memory is without doubt a necessary condition for our style of cognition, it's harder to argue that it's a sufficient one.

It is, of course, vastly entertaining to speculate about just what physical constituent it is that makes the human brain, and thus our cognition, truly unique. But, at least for the moment, there is a very strict limit to what we can directly observe in the historical record. What is more, for all the advances that have been made in visualizing exactly what is going on in the brain when it undertakes various activities, we are still at a loss to explain how all that activity is integrated to form our consciousness of ourselves and of the world around us.

We have seen that the brain is a living, changing tissue, constantly rewiring itself in response to experience. It is not beyond possibility that changing cultural stimuli, as human life has become more complex, have contributed to across-the-board changes in the modern human brain. Still, some new biological ingredient must condition all the responses the modern brain makes throughout life. This new physical element was most plausibly acquired about two hundred thousand years ago, along with all of the other anatomical correlates of our modernity. But we still can't be sure just what it was.

EPILOGUE

The ultimate origins of our remarkable brains are remote indeed in time, and they recede in a near-infinite succession of levels of complexity. The physics that underwrites brain function at the most basic of those levels goes all the way back to the origin of our universe; the chemistry that characterizes brains as living tissues originated not so very long after earth itself; and the first intimations of something special about what were to become brain cells goes back at least to our common ancestor with sponges. Structures that members (like us) of a vertebrate species might intuitively recognize as a sort of brain appeared a whopping five hundred million years ago, and among the huge variety of vertebrates that share the planet with us today we can identify numerous living forms with brains or brainlike structures that very roughly approximate the multiple "stages" of neural complexity through which our own ancestors must have passed. In terms of understanding the overall architecture and workings of the brain, looking at the Tree of Life of which we form an integral part allows us to speculate with some confidence, and in great detail, about its long, messy, accretionary history. But we are still left fairly helpless in contemplating just how we acquired the brain's greatest gift to us: our unique form of consciousness.

After examining everything we know about our long series of ancestors, whether hypothetical or matched by some living form, or known in the fossil record, we still have no idea exactly how our symbolic consciousness came about or how it is enabled by the particular wiring of our brains. We know approximately when the transition to symbolic thinking occurred, and we know quite a lot about the long sequence of events leading up to it. We can even concoct historical scenarios of its origin (chapter 10). But the great functional mystery remains. We are nowhere near to comprehending how all those electrochemical signals in our brains add up to what we (or any other organisms, of whatever complexity) encounter daily as our consciousness, which sums out as our individual, subjective experience of the world. As a result, we have no idea exactly what our brains are doing that is different from what happens in the brains of other cognitively complex forms that do not share the symbolic facility. All we know is that the human brain is differentially enlarged in certain regions compared with its closest living and fossil relatives; that some of its tissues also show architectonic differences from their equivalents in its closest living cousins; and that in some respects, at least, we human beings process the information that comes in from our senses in a way that is profoundly different from what we see in any other organisms, even including those closest relatives.

But whatever the biological underpinnings of our cognitive uniquenesses may be, the results of those differences are dramatic. So distinctively do we think, indeed, that there is absolutely nothing in what we know about the biological and behavioral trends that preceded our "transition" to symbolic thought that would ever have allowed a bystander to predict that this transition was going to happen. Indeed, human beings deal with what they perceive of the world around them in such an entirely unprecedented manner that, as we've already remarked, the only reason for thinking that any living form *could* ever have acquired symbolic consciousness is that our species so clearly *did*.

As the only hominid species in the world today, we *Homo sapiens* find it very tempting to reconstruct our history by projecting ourselves back into the remote past, thus viewing our evolutionary precursors as increasingly inferior versions of ourselves as we go back in time. After all, we are not in conflict with any other hominid species right now, and there are no other actors in the evolutionary play to threaten us, or that we need in any way to measure ourselves against, other than to determine our place in the Tree of Life. It is because of these unusual circumstances that we have the luxury of choosing to see ourselves from the perspective of progressive improvement over time, a viewpoint that is particularly attractive to us for reasons that go well beyond simple ego gratification. As members of a storytelling species, we readily respond to any explanation of our origins that uses a vocabulary we can directly relate to our own experience. We don't have to step outside our unique human mindset to view our predecessors as to varying degrees less smart than us; all we have to do from this perspective is just to knock off a few IQ points. In stark contrast, it's almost impossible for us to imagine the qualitatively alien cognitive states that were almost certainly experienced by our relatives at earlier stages of human evolutionary history. What is more, we just happen to live in an unprecedented time of rapid and continuous technological advance, which positively invites progressivist notions of the past. Once again, we are lured by the familiar to extrapolate our own personal experience back into our histories.

Still, seductive as the resulting view of ourselves as provisionally perfected entities may be, this book reveals not only that the human brain is hardly a perfect product but also that the mistaken notion of evolution as a process of inexorable improvement under the benevolent hand of natural selection hugely distorts our understanding of our place in nature. Every living species, no matter how simple it may look from our point of view, is in reality an endpoint, a terminal twig on the Tree of Life. Living and surviving in the moment, no species is "going" anywhere. Life is, after all, about getting by today, for if you don't succeed in the moment, there is no future. Evolutionary trends are something we only ever see in retrospect; and in most cases they result from the differential success of entire species, rather than from the honing of lineages over long stretches of time. For all its unusual qualities, *Homo sapiens* is just another of those myriad endpoints on the Tree of Life—and so was each of its predecessors.

As we've seen, evolution cannot invariably, if ever, be the process of optimization implied by the gradual-change model in which species become ever more perfectly adapted to their environments. For one thing, environments tend to change abruptly, on short timescales to which natural selection cannot possibly respond. And if your environment does change suddenly, you are much more likely to move somewhere else more congenial or, if this is impossible, to become extinct than you are to stand still and change on the spot. Indeed, the less tightly you are adapted to a particular environment, the more flexible you are and the less likely it is that the environment will sock you hard when it changes. There is a price to be paid for optimization, as we see in higher extinction rates among specialist species than among generalists.

The "messy" human brain is a wonderful example of the kind of untidy history that this view of evolution reflects. We have already mentioned that Gary Marcus accurately described our brains as "kluges": inelegant but effective solutions to a problem. As the products of a long evolutionary process with many zigs and zags, involving ancestors who lived in circumstances that were hugely different from our own, our brains are splendidly jury-rigged affairs that no engineer would ever have designed. Indeed, to a large degree it is because the human brain has *not* been optimized for anything that it is the hugely creative, and simultaneously both logical and irrational, organ that it is.

So we need to disengage a bit. We need to look at ourselves not so much as the perfected outcome of a long process of improvement over the eons but as the lucky winners of a long evolutionary tournament in which the rules just happened to suit us. Indeed, a mere one hundred thousand years ago our species' situation was very different and would have mandated a different perspective on human origins. At that time we shared the world with several other hominid species, as all our predecessors had, and *Homo sapiens* doesn't appear at this point to have behaved significantly differently from any of its contemporaries. But once our species had acquired— immediately, at least, as the result of a purely *cultural* stimulus—the ability to reason symbolically, all bets were off. An entirely novel cognitive entity was on the scene, and whatever the exact details may have been of the demise of the Neanderthals and *Homo erectus* (and any other hominids of the time), the fact is that they proved incapable of competing effectively with symbolic *Homo sapiens*. Those other hominids of the last Ice Age, the unfortunates who were forced to deal with the new phenomenon of cognitively modern and rapidly dispersing *Homo sapiens*, were surely sophisticated and subtle interactors with their environments. But they were the last of the "old-style" hominids. Cognitively modern *Homo sapiens* was clearly at a relative advantage, at the very least in the domains of planning and the ability to intensify exploitation of the environment. And whether or not there was ever any direct conflict, the economic advantage those cognitive abilities supplied simply proved too great for hominid rivals to resist.

We all know that human behaviors are sometimes bizarre, and in recent years the discipline of "evolutionary psychology" has prospered by producing a string of attractive logical explanations for our behavioral failings. The basic idea, born of the notion of evolution as a simple process of fine-tuning, is that our brains are somehow out of kilter with the modern world because they were specifically formed by natural selection to cope with an "environment of evolutionary adaptedness" that no longer exists. We act inappropriately because we find ourselves in a modern urbanized world even as we are still saddled with a mindset that is fine-tuned to the exigencies of living as hunter-gatherers in tiny groups, thinly scattered over Ice Age environments. To evolutionary psychologists, it is this disconnect between the world we live in and the world we are adapted to occupy that accounts for the weird ways in which we sometimes behave. This is a great story, and one with a strong innate appeal to members of a storytelling species. But alas—or, rather, very happily—it is just plain wrong. The evolutionary psychologists have played a valuable role in documenting recurring patterns among our behaviors that most certainly reflect the common structure of the neural equipment residing in the heads of all cognitively normal human beings. But blaming our foibles on now-inappropriately fine-tuned brains is entirely missing the point. Our minds never were fine-tuned. Instead, our brains and the occasionally bizarre behaviors they produce are the product of a lengthy and supremely untidy history: a history that has accidentally resulted in a splendidly eccentric and creative product. Our stupidities and imperfections are no more than the flip side of our brilliance.

One of the by-products of our symbolic cognition is the human penchant for storytelling that we have already emphasized. Storytelling is deeply ingrained in our psyches, and it is one of the key characteristics that keeps boredom (another of those by-products) at bay. Science is, of course, a form of storytelling, but it is a special kind of storytelling—one that demands, as we have seen, that the stories we tell should be testable against what we can observe in nature. "Just-so" stories, however entertaining, need not be constrained by reality and do not qualify as science. This means we should be particularly wary of the elaborate adaptive explanations scientists often concoct to explain our species' manifold imperfections. The temptation to come up with special explanations of this kind is made irresistible by the assumption that evolution and improvement go hand in hand. And even though the assumption itself is misleading, it's nonetheless undisputable that, for all its exceptional intelligence and abilities, the human species is notably flawed: from its physical tendencies for flat feet, hernias, and slipped discs all the way to the stupid behaviors we all exhibit, no matter how smart we are. We have irrational beliefs; we are horribly bad at judging risk (especially—and very dangerously for the species and planet—long-term risk); we are prone to extreme cognitive dissonance; and we act self-destructively in more ways than we could ever specify here. All this is undeniably the case, and it is something we consciously need to address in our individual and corporate lives. Nonetheless, our bizarre condition is much better explained by the inherent general "messiness" of evolution than it is by special pleading in particular cases. Indeed, we humans are precisely the kind of product you would expect from an opportunistic process that is not in the least concerned with optimization.

Of course, for all our deficiencies we may justifiably describe ourselves as the smartest beings on the planet, able to manipulate information in our heads and, thanks to our wonderfully dexterous hands, to implement our conclusions to affect the outside world. But we always need to remember that "smart" is a relative term. In many areas of brain function, there are creatures who do the job better than us. In a profound sense, every terminal twig on that great Tree of Life is equally evolved. Avian raptors are able to spot movements far away in the landscape that our eyes could never detect, and a dog's experience of the olfactory world is far richer than our own. Indeed, it's fair to say that in a very real sense dogs and we live in entirely different (if overlapping) worlds, in which the two species respond to very different sets of cues. The human way of experiencing the world is only one of very many possible, and this brings us back to the importance of metaphor. When comparing types, or even degrees, of intelligence, we are limited to doing so at the level of metaphor, which is a purely heuristic device to help us understand where we fit into the world. Describing ourselves as the smartest creature on the planet is just to say that we do what we do better than any other creature does it. In fact, for better or for worse, we are the *only* creature that does it.

When we go out and talk to audiences about human evolution, one question always crops up. And it's a very good one, too. Questioners point out that in the history of the genus *Homo* the average brain size of hominid species dramatically increased with time—implying that hominids were consistently getting smarter (in that metaphorical sense, we'll emphasize). And they want to know whether, as a result of this strong and undeniable past trend, it is reasonable to expect that in the future our brain size and hence intelligence will continue to increase. Well, there are two parts to the answer here. One refers to the *quality* of the intelligence involved. Certainly, the intuitive-style intelligence of our precursors became more complex and subtle with time, as reflected in the kinds of archaeological evidence of their lives those ancient hominids left behind. But there was a major inflection point at the very recent origin of symbolic thought, when the human brain evidently began to think in an entirely new way. And at this point in our own history we are still right at the beginning of the new curve; we really have no track record from which to extrapolate. What's more, if anything, human brains may even have become marginally smaller on average since Cro-Magnon times, perhaps indicating that the new way of processing information demands no more bulk of neural tissue than existed before—and will not in future.

The other aspect of what we might expect in the future relates to the nature of evolution itself. Our species evolved in a world in which hominid populations were thinly scattered over vast tracts of the earth's surface, continually buffeted by the vagaries of rapidly changing Ice Age climates. These are the ideal conditions in which to expect that genetic novelties randomly arising within populations should become fixed. Large populations have such great genetic inertia that the probabilities of any significant innovations being incorporated are very low indeed—just as we saw in the coin-flipping example we quoted in chapter 1. In large populations, gene frequencies will always slosh around a bit. But the key ingredients for meaningful biological change are two: small effective population sizes and a high degree of isolation among those populations.

Conditions ideally conducive to evolutionary change clearly held sway during the early years of *Homo sapiens*, when our predecessors were still hunters and gatherers, roaming in relatively low numbers over vast territories. But once humans began to live sedentary lives in (unprecedented) response to the climatic crisis at the end of the last Ice Age, everything changed. Planting crops demanded labor, and the reproductive limits imposed on human populations by the limited ability of mothers to carry more than one or two slowly developing children around were lifted. Populations skyrocketed and have been increasing ever since. Today we are living in a very crowded world, in which virtually every inhabitable stretch of the earth's surface has been occupied. What's more, individuals within that huge worldwide population are more mobile than ever before. Never have conditions been *less* propitious for the incorporation into the human species of meaningful biological novelty. Certainly, the human genetic deck is being vigorously reshuffled, as formerly isolated populations mingle as never before. But the cards remain basically the same: significant change is simply not in the offing. Imperfect we may be, but we cannot hope for evolution to rescue us from our imperfections.

Of course, there remains a possibility—an extremely remote one, we hope—that our very imperfections may change the rules by fragmenting and reducing our huge and expanding population once more, either through deliberate action or by inaction in the face of threats that are already looming. But, short of an apocalyptic event (anthropogenic or otherwise) that most of us would vastly prefer not to see, we would do better to recognize our cognitive imperfections and to use the intelligence we have to devise ways of coping with them.

TIME LINE

Million Years Ago	Event
15,000–13,700	Big Bang, and chemical sequelae
4,500	Earth forms
4,000	Most likely origin of the RNA world
3,500	Cellular life evolves on earth
3,000–2,500	Photosynthesis evolves and earth's atmosphere becomes oxygenic
1,500	Ancestor of Archaean and Eukaryotic cells
1,000	Multicellular organisms arise on earth
700	Ancestor of simple animals (Porifera, Cnidaria, Placozoa, Ctenophora, primitive Bilateria)
600	Simple animals begin to diverge into major groups
570	Common ancestor of Arthropods (insects, myriapods, onychophora, crustaceans, etc.)
550	Complex animals begin to diverge (Cambrian Explosion)
500	Vertebrates begin to diverge
475	Plants begin to diverge
400	Insects and Seed Plants diversify
360	Amphibian ancestor
300	Common ancestor of Reptiles and Birds

200	Origin of Mammals
150	Origin of Birds
~130	Origin of Flowering Plants
65	Extinction of Dinosaurs, expansion of Mammals
64–60	Common ancestor of Primates
35–30	Common ancestor of Higher Primates
8–6	Chimpanzee and human lineages diverge
4.2	Genus *Australopithecus* established
2.5	First manufacture of stone tools
2.0	Genus *Homo* arises; trend to increasing brain size among hominids begins
1.5	First deliberately shaped stone tools
0.8	First control of fire in hearths
0.4	First constructed dwellings
<0.2	Anatomically modern *Homo sapiens* in Africa
~0.1	First stirrings of symbolic behaviors in Africa
~0.035	First cave art
0.001	Domestication of plants and animals begins
0.0005	Earliest writing systems

GLOSSARY

acetylcholinesterase: An enzyme that degrades the small neurotransmitter molecule acetylcholine.

Acheulean: The stone tool-making industry characterized by the teardrop-shaped and symmetrically shaped hand axes that began to be produced in Africa about 1.5 million years ago. Named for Saint-Acheul, in France, where they were first identified.

action potential: Action potentials occur when nerve cells fire and are characterized by a change in electrical charge or potential on the surface of a nerve or muscle cell. Action potentials are short-lived and result when the membrane of a neural cell opens to allow positive ions to pass to the inside of the cell, and negative ions to the outside. Changes in charge of nerve cells are implemented by action potentials. Nerve impulses are the result of series of action potentials moving down the length of a nerve cell.

ADAR (adenosine deaminase acting on RNA): An enzyme that regulates the process of RNA editing. RNA editing has been shown to be an important process for regulating the expression of genes.

agonist: A chemical or molecular substance that can bind to a specific receptor protein and, as a result, implement a cellular response. Three agonists that we discuss in this book are AMPA, NMDA, and kainite, which influence receptors that glutamate also interacts with. All three of these chemicals are analogues of glutamate and therefore can interfere with the normal activity of glutamate at nerve synapses. The receptors that these agonists interact with are all ionotropic at the synapse.

amygdala: A bilaterally symmetric paired structure in the inner brain considered part of the limbic system. This region of the brain is thought to be important in processing information that is uncomfortable to the organism, such as fear or unhappiness.

anatomy of a neuron: Neural cells are variable in their shapes and sizes. The soma, or central part of the neuron, takes up the bulk of the neuron and contains the nucleus of the cell. Dendrites of neurons vary in diameter and are cellular extensions of the neural cell that give the cell its treelike appearance. Most of the action potential input into the cell occurs through the dendrites. The axon is a cablelike extension of the cell and is the part of the nerve cell that carries the action potential (or nerve impulse) away from the soma. Axons make connections with other neural cells through synapses.

angular gyrus: Association area in the parietal lobe (near the junction of the temporal and occipital areas) identified by Norman Geschwind as a key area in which cortico-cortical connections can be made without limbic intermediation—and which is thus possibly connected to object naming and language. Unfortunately, this area is hard to visualize on fossil endocasts.

antibody staining: A procedure used by experimental biologists to localize, to the exact position of individual cells, the production of a gene product. Genes produce proteins, and proteins can be recognized by antibodies. Scientists have learned to attach marker molecules such as "fluorescent beacons" to antibodies. When the beacons are allowed to bind to tissues in which a protein of interest is produced, the fluorescent beacon indicates exactly where in the tissue the protein resides.

aphasias: Impairments of language ability that include several syndromes such as verbal dyspraxia, which disturbs coordination in the movements of the speech production system.

Arthropoda: Including Chelicerata, Crustacea, Myriapoda, and Hexapoda, arthropods are animals without vertebrae but having an external skeleton (exoskeleton) and segmented bodies with jointed appendages. Arthropods belong to the phylum Arthropoda, of which Chelicerata, Crustacea, Myriapoda, and Hexapoda are subphyla. Chelicerata includes spiders, horseshoe crabs, and scorpions. Myriapoda includes centipedes and millipedes. Crustacea includes organisms like shrimp, crabs, and lobster, and Hexapoda includes insects. Close relatives of the arthropods are the Onychophora and the Tardigrada (water bears); the three groups together make up what taxonomists call the Panarthropoda.

Australopithecus: Genus that includes the earliest well-documented extinct human relatives. It consists of several African species known in the period between about 4 and 1.5 million years ago. These early hominids were upright bipeds when on the ground but they were of very modest stature, with short legs, and they retained features particularly of the upper body that suggest retained agility in the trees. It was probably early hominids of this general kind that started making stone tools some 2.5 million years ago. Several species within this genus developed massive chewing dentitions and are often alternatively allocated to the genus *Paranthropus*.

auxin: A category of small molecules also known as plant hormones. Auxins are plant growth substances and are important in development.

bacteriorhodopsin: A transmembrane protein found most prominently in the archaeans known as Hallobacteria. The protein's major role is as a light-driven proton pump. The protein uses light to move protons across the membrane to the outside of the cell, producing a gradient that is converted into chemical energy.

Bilateria: A category of metazoan animals with bilateral symmetry. There are two great lineages within the Bilateria, defined by the kind of cleavage that occurs during embryonic development. The protostomes cleave their embryos in a spiral fashion, and the deuterostomes cleave in a radial fashion. The protostomes are in turn divided into two major lineages, the Ecdysozoa (animals that molt like insects) and Lophotrochozoa (animals with a specific kind of feeding apparatus, like mollusks).

brain imaging techniques: Several novel methods of imaging the brain have been developed in recent decades. These include functional magnetic resonance imaging (fMRI), positron emission tomography (PET), magnetoencephalography (MEG), and diffusion tensor imaging (DTI). All of these methods use some by-product of brain activity detectable by a sensor, coupled with high-powered computing, to reconstruct where in the brain an activity occurs. fMRI uses the change in blood oxygen as a means to detect functionality in the brain. The difference between traditional MRI and fMRI is that MRI is done on static resting subjects, whereas fMRI is accomplished on subjects performing specific brain-related tasks. PET requires the introduction of a radionuclide connected to a biologically active molecule that is involved in brain function. PET then detects the activity of the active molecule via the release of radioactivity. MEG records the magnetic field produced by brain activity. DTI measures the how water and neurotransmitters flow, or diffuse, through the brain. The diffusion patterns reveal pathways for neural connections in the brain.

Broca's area: A small patch of cortex within the left frontal lobe, with motor functions that include control of the speech apparatus.

caudate nucleus: A structure in the brain within the basal ganglia. It is involved in learning and memory.

cerebral fissures: The wrinkles on the surface of the brain due to the folding of the brain's cortex to fit within the skull. The major wrinkles are generally thought to delineate important functional areas.

choanocyte chamber: Choanocytes are a cell-type found in sponges that have flagellae. These cells line cavities, known as choanocyte chambers, in the sponge body.

cingulate cortex: Considered part of the limbic system, this part of the brain coordinates information from the thalamus and neocortex and is important in processing emotion, learning, and memory.

coelom: A cavity filled with fluid, found in some animal body plans. The cells that line the coelom develop from the mesodermal layer of the embryo. A coelomate is an animal with a coelom.

cortisol: A steroid hormone secreted into the bloodstream by the adrenal gland. It is a small molecule that acts as an agonist in many functions of the brain and the body. It is also known as a stress neurotransmitter.

cryptomonads: A group of single-celled eukaryotes forming part of the algae most commonly found in freshwater. Most cryptomonads have intracellular organelles called plastids and are characterized by the presence of extrusomes. These are small spiral structures that are kept under tension by the cell. When disturbed, these extrusomes expand, causing the cell to move away from the disturbance.

cyanobacteria: An ancient group of Bacteria that uses photosynthesis to obtain energy from the environment. Cyanobacteria are thought to have been responsible for converting the early reducing atmosphere of the earth into an oxidizing atmosphere and are thought to have originated about two billion years ago. Cyanobacteria are also the lineage of bacteria that gave rise to chloroplasts in plants, through the engulfing of ancient cyanobacteria by the eukaryotic lineage that eventually gave rise to plants.

cycads: Higher plants that bear seeds are divided into two great lineages: the angiosperms (flowering plants) and the gymnosperms (nonflowering). Cycads are a member of the gymnosperm group, as are ginkgos.

diencephalon: Along with the telencephalon, this region of the vertebrate brain makes up the forebrain. It abuts the mesencephalon (midbrain) and lies between the midbrain and the telencephalon. It consists primarily of the thalamus, the hypothalamus, and the pretectum.

Dmanisi: The 1.8-million-year-old site in the Caucasus (Republic of Georgia) where the earliest non-African hominids have been identified.

domains of life: The three major domains of life are Eukarya, Bacteria, and Archaea. Currently scientists place all forms of life on the planet into one of these three domains. Eukaryotes are organisms that are best characterized by possessing a membrane-bound nucleus. Archaea are characterized by the reverse stereochemistry of glycerol, while Bacteria have genes lacking introns.

dopamine: An endogenous small molecule that is a neurotransmitter released in specific parts of the brain. It is primarily important in the regulation of the hormone prolactin, but it also has a role in the way the brain perceives motivation, punishment, and reward. It is also important in voluntary movement and cognition.

encephalization quotient (EQ): A figure that measures the size of an animal's brain compared to its body size. Technically, it is the ratio between actual brain mass and the size statistically predicted for an animal of similar body mass, although techniques for making this estimate differ.

endocast: The brain develops in intimate relationship to the inside of the braincase, which thus preserves a fairly accurate record of its external shape. An impression (either natural or artificial) of the inside of a fossil braincase will thus quite accurately reproduce the size and form of the brain the vault had contained. The scientists who study the brain endocasts of extinct species are known as paleoneurologists, and they are usually fairly cautious in the statements they make because what the brain does is more a function of its interior connections than of its precise external form.

endorphins: Small endogenous peptides that are produced by the pituitary gland and the hypothalamus. They act as neurotransmitters and are involved in a large number of brain functions, including love, orgasm, and general excitement. Opiates such as heroin and morphine resemble endorphins and hence can interfere with the normal functioning of endogenous endorphins.

evolutionary synthesis: The body of work accomplished by evolutionary biologists in the twentieth century that amalgamated Darwinian evolutionary thought with genetics and modern biology. The mathematical genetic phase of the synthesis was accomplished by Sewall Wright, R. A. Fisher, and J. B. S. Haldane. The empirical phase of the synthesis was implemented primarily by three biologists, Ernst Mayr, Theodosius Dobzhansky, and George Gaylord Simpson. In its "hardened" form, the synthesis more or less equates evolution with gradual change.

exaptation: A morphological or physiological trait that is used in a context unlike that in which it was originally acquired.

extremophiles: Organisms like Bacteria and Archaea that have adapted to extreme environments, like highly acidic or basic ones, extremely hot or cold environments, or environments under high pressure.

fossil: Technically, any and all traces of past life; most commonly, the mineralized bones and teeth of ancient animals that happen to have been preserved, usually in fragmentary state.

14-3-3 proteins: A family of highly conserved molecules that regulate gene expression. They are found in all eukaryotic cells and are multifunctional in that they can bind to and regulate a wide array of signaling proteins.

FOXP2: A protein, found in animals, that is thought to be necessary for some aspects of speech and therefore involved in language production.

G-proteins: Also known as guanine nucleotide binding proteins. This large family of proteins assists in transmitting signals from outside the cell to the inside. In doing so they alter the state of the inside of the cell. Hormones and neurotransmitters are the source of many of the signals that G-proteins communicate form.

G-protein complex interactions: G-proteins interact with guanosine tri- and di-phosphates (GTP and GDP, respectively), common molecules that are important for the metabolism of the cell. The G-protein complex is turned on when it is complexed with GTP and turned off when complexed with GDP. But to get the GTPs and GDPs to interact with the G-proteins, something from the outside of the cell, such as a neurotransmitter receptor, has to complex with the GPCR itself.

G-protein-coupled receptors (GPCRs): This large family of proteins possess seven membrane-traversing domains. As their name implies, they couple to G-proteins. Their function is to interact with and hence "sense" molecules outside cells. These then activate signal transduction pathways on the inside of the cell that in turn trigger a cellular response.

gamma aminobutyric acid (GABA): A nonalpha amino acid that serves as a ligand in the nervous system. GABA is the major inhibitory neurotransmitter of the central nervous system in mammals.

gene expression: Genes are made of DNA that needs to be transcribed into messenger RNA and then translated into amino acid chains to produce proteins. Both transcription and translation are the products of gene expression, and scientists use both messenger RNA and protein detection methods to observe gene expression.

glial cell: The nervous system is made up of two basic types of cells, neural cells and glial cells. Glial cells make up about half of the cells in the nervous system and have classically been thought to provide mechanical support and insulation and protection of neurons. Recent data suggest that glial cells themselves may transmit impulses and are capable of communicating with other cells in the brain.

globus pallidus: A subcortical region of the brain associated with the thalamus. It is part of the basal ganglia and has complex relationships with other parts of the inner brain (the striatum and substantia nigra). It is involved in processing and regulating motor functions.

glomeruli: Nerve cell structures found in the olfactory bulb of the vertebrate brain. A single glomerulus is a globular tangle of axons emanating from the olfactory receptor neurons.

gross structure of the vertebrate brain: At the coarsest level the brain can be divided into four regions, going from posterior to anterior—spinal cord, hindbrain, midbrain, and forebrain. The spinal cord is responsible for very primitive automatic responses to the environment. The hindbrain includes the cerebellum, pons, and medulla oblongata. The midbrain includes the substantia nigra, and the forebrain includes the limbic system, basal ganglia, and cortex.

habituation: A form of nonassociative learning that occurs after repeated exposure to an external stimulus over a long period of time. An organism will decrease its behavioral response to a stimulus if no harm is done as a result of the stimulus.

heterochrony: The change in developmental timing that leads to changes in size or shape of organisms, organs, or tissues. By changing the timing of a developmental event, wholesale changes in shape or size can sometimes be implemented in an organism.

hippocampus: A bilaterally symmetrical paired region of the brain commonly associated with the limbic system. It gets its name from its shape, which is supposed to look something like a seahorse. The hippocampus lies close to the midline of the brain, and it is responsible for processing and consolidation of new information and hence is involved in short- and long-term memory.

homeotic genes: A category of genes involved in the development process that were given the name because they produce transformations of one body part or region into another. Both plants and animals have homeotic genes. The homeotic complex is a subset of homeotic genes (that is, not all homeotic genes are HomC genes). The HomC genes control segmental identity in animals and have within them a conserved sixty-one amino acid motif called a homeobox.

hominids/hominins: Members of the group containing human beings, plus all of their extinct relatives that were not closer to the living great apes. It is becoming increasingly common for paleoanthropologists to recognize this group at the subfamily level (Homininae) rather than at the family level (Hominidae), but in practice it doesn't make much difference. We use "hominid" here largely out of habit.

Homo: The genus to which modern humans belong. It is poorly defined, with some fragmentary fossils over 2 million years old often allocated to it. However, the characteristic tall, slender, striding *Homo* body form was only certainly established rather later than this, by about 1.8 to 1.6 million years ago. Following this the genus diversified considerably and started on a trajectory of brain size enlargement, apparently independently in several lineages. It is highly unusual for our species, *Homo sapiens*, to be the lone hominid on the planet.

homoplasy: A biological convergence that results in two unrelated organisms appearing to share the same trait. Homoplasies can be misleading: if you were to conclude that birds and bats are more closely related to each other than to a human because both have wings, you would be deeply mistaken.

ion channels: Proteins, embedded in the cell membrane, that form passageways or pores that establish and control the electrical gradient across the cell membrane of living cells. They control the flow of ions into and out of the cell. Ion channels are classified by the way they open and close in a process called gating. There are three major gating categories: ion gating, ligand gating, and other more complicated (but less common) ways. Ion-gated channels are controlled by ion concentrations inside and outside of the cell. Ligand-gated channels use small molecules (ligands) that bind to receptor molecules, embedded in the membrane, that in turn implement the opening of the channel.

ionotropic receptor: A receptor protein that is fast-acting and serves as an ion channel gate. GABA acts as a ligand for such receptors, and upon the binding of GABA to this kind of receptor, the gated channel opens.

kluge: A clumsy or inelegant—but effective—solution to a problem. This term has recently been applied to the human brain by New York University psychologist Gary Marcus.

language: A form of communication unique to modern humans (and usually spoken) that depends on creating a finite vocabulary of symbols (spoken or signed) that can be combined according to rules to make an infinitely large number of statements about the world.

L-DOPA: Also known as levadopa, this small molecule is the precursor to several other compounds called catecholamines. These include the neurotransmitters dopamine, norepinephrine, and epinephrine.

ligand: An ion, a molecule, or a more complex group of molecules that binds to chemical entities such as proteins (such as neurotransmitter receptors) to form larger molecules that usually trigger a change in state of the chemical entity.

lipids: Biochemical compounds that are best characterized as fat or oil. Lipids will dissolve in alcohol but will remain intact in water.

Middle Stone Age (MSA): The African cultural phase (lasting from about two hundred thousand to thirty thousand years ago) during which the earliest stirrings of the symbolic spirit can be detected.

mechanoreceptor: A receptor molecule that responds to mechanical pressure or to being distorted. Most mechanoreceptors reside on the skin.

memory, types of: Memory is a complex entity that can be categorized in many ways. The most common way of categorizing memory is into declarative and procedural memory, which are roughly equivalent to explicit and implicit memory, respectively. Procedural (implicit) memories are those that aid in doing tasks that don't take conscious awareness, like riding a bike or brushing teeth. Declarative (explicit) memories are those that we incorporate into our conscious understanding of the present. Declarative memory can be divided further into episodic and semantic memory. Episodic memories are those that pertain to names, places, and experiences of the past, and semantic memories are those that pertain to definitions or concepts. Memory can also be categorized as long-term or short-term. These two categories refer to storage, as long-term memories are those that the brain deems worthy of long-term maintenance and short-term memories are those that the brain will retain for only a short period of time.

metabotropic receptors: These receptors act quite differently than ionotropic receptors by using second-messenger proteins. They implement a slow response and use a category of intracellular proteins called G-coupled proteins.

metazoans: A major group of organisms that embraces the multiple-celled animals. The phylogeny of the Metazoa is controversial, but the group includes the following major animal lineages: Porifera (sponges), Cnidaria (cubozoans, jellyfish, corals, and hydra), Placozoa, Ctenophora (comb jellies), and Bilateria like us.

microcephaly: An anatomical disorder whereby the size of the brain is drastically decreased. Genetic research has recently localized several genes involved in the syndrome.

miRNA-induced silencing complex (RISC): A complex of proteins produced by a cell that bind to each other to process small interfering RNA (siRNA).

Mousterian: Stone-working industry usually associated with Neanderthals (but occasionally with early modern humans).

mutualist: A type of interspecies interaction that is beneficial to both species involved in the interaction.

neocortex: Part of the cerebral cortex that has expanded in mammals. It exists as six cell layers (numbered I through VI) and in mammals is involved in higher functions such as decision-making, determining spatial location, conscious thought, sensory perception, and (in humans) language.

neuroeconomics: The study of the neural aspects of human interactions involving economic decisions. This new field incorporates economic theory, psychology, and neurobiology to study the process of decision-making.

neuromodulation: A neural process whereby multiple categories of neurotransmitters, released in tight concert, affect the regulation of several neurons or groups of neurons.

neurotransmitter: Most commonly neurotransmitters are small molecules, like glutamate, that transmit signals from a presynaptic cell across a synapse to a postsynaptic cell. They are packaged in what are called synaptic vesicles in the presynaptic cell and are released into the synaptic region between two cells that forms the synapse. On release, they bind to receptors on the surface of the postsynaptic cell. The release and subsequent binding of neurotransmitters is usually accompanied by action potentials.

nociception: The neural processing of noxious stimuli.

norepinephrine: A small molecule that acts as a hormone in the body and as a neurotransmitter in the brain. It is involved in reactions to stress.

nucleic acids: Complex biochemical compounds found in all living cells and also in viruses. They are "complex" because they are composed of purines, pyrimidines, carbohydrates, and phosphoric acid. The two most common nucleic acids in living cells and viruses are deoxyribonucleic acid (DNA) and ribonucleic acid (RNA).

Oldowan: The name generally applied to the earliest kind of stone tool, which consisted of simple sharp flakes struck from a core. The name comes from Tanzania's Olduvai Gorge, where such tools were originally identified. First known from about 2.5 million years ago, such tools only began to be supplanted by more sophisticated forms a million years later.

Onychophora: Known commonly as velvet worms, these invertebrates are members of a small phylum called Protracheata. Thought to be close relatives of arthropods, they are placed in a larger taxonomic unit, Panarthropoda, along with Arthropoda and Tardigrada.

opioid receptors: A family of transmembrane G-protein-coupled receptors (GPCRs) that are bound by the opioid ligands. There are four major kinds of opioid receptors: kappa, mu, delta, and nociceptin.

opsins: These proteins come in two kinds, Type I and Type II. Type I opsins are found in bacteria and have many different names such as bacteriorhodopsin and proteorhodopsin. Type II opsins are found in all animals except sponges. They are even found in placozoans and all cnidarians examined to date. Most Type II opsins have a role in vision, but some, oddly enough, have a role in determining circadian rhythm, or the inborn rhythms of approximately twenty-four hours that regulate a lot of the physiology of organisms. A small category of opsins also has a role in eye pupil constriction. The existence of Type I and Type II opsins in such different groups, together with their absence in plants, most fungi, and many protists, brings us to the most interesting statement we can make about opsins: they evolved twice!

osmosis: The movement of solvent molecules across a selectively permeable membrane such that the concentration of the molecules equalizes on both sides of the membrane.

outgroup: When systematists are deciphering the relationships of a group of organisms, they use an outgroup to "polarize" those relationships. An outgroup is simply defined as a group of organisms that does not belong to the group being systematized—though it will usually be quite closely related. For example, if a bird and a bat and a human are examined with a phylogeny in mind, a good outgroup would be a fish (which is not in the bird group or the mammal group).

paedomorphosis, or neoteny: A heterochronic process whereby developmental timing is altered so that a juvenile shape is retained into later life stages. The result is an individual of juvenile appearance that is reproductively mature. An excellent example of paedomorphosis is the axolotl, a salamander that retains the tadpole morphology into its adult stage.

phospholipid: A lipid with at least one phosphate connected to it. Phospholipids have a "head" that is is *hydrophilic* (water-loving) and a "tail" that is *hydrophobic* (water-hating). Phospholipids are the major component of cell membranes.

phosphorylation: The addition of phosphate or phosphate groups to a target protein. Such addition alters the function of the target protein in one of three ways—changing enzyme activity, changing cellular location, or causing a change in association of the phosphorylated protein with other proteins.

potentiation, long-term (LTP): A persisting increase in the strength of a synapse after repeated stimulation.

prepared-core: Method of making stone tools whereby a core was laboriously prepared using multiple blows until a more or less finished flake (or a succession of them) could be struck off with one blow (or a succession of them).

Prokarya: The classical grouping of organisms without membrane-bound nuclei (Bacteria and Archaea). While used for decades, this grouping is "unnatural" because Archaea and Bacteria are not each other's closest relatives.

promoter regions: Regions of a gene that regulate the expression of the gene. Genes have "coding regions" that are translated into protein as well as regions that are not translated (called "untranslated regions," or UTRs). UTRs often implement the regulation of the gene by either promoting or impeding its expression.

proprioceptors: Sensors that give information about the position of various body parts in space.

protein kinases (PK): Enzymes that implement the chemical addition of phosphate groups to other proteins (in a process called phosphorylation). More than five hundred genes in the human genome have protein kinase activity.

proteorhodopsin: A protein that is homologous to bacteriorhodopsin but is found in marine bacterioplankton and marine eukaryotes.

putamen: A part of the basal ganglia. Along with the caudate nucleus, it makes up the striatum. This part of the brain is important in regulating movements and learned behaviors.

random drift: Changes in allele frequencies resulting from sampling error. Drift is a stochastic process that often leads to counterintuitive results.

resistance: Opposition to the passage of an electrical current. In the context of nerve cells, resistance is opposition to the passage of action potentials. The opposite of resistance is conductance, or the enhancing of the passage of electric current.

ribosomes: Cellular structures composed of an RNA backbone augmented with proteins. There are two subunits to a ribosome (large and small). The ribosome is the site of protein synthesis in the cell.

scala naturae: A discredited way to classify and categorize life on the planet, initially devised by Aristotle. In this classification, life is arranged from the less complex to the more complex, forcing a linear classification with humans at the top.

second messenger system: The system by which second-messenger molecules relay signals from the surface of cells, via receptors that reside on the cell surface, to target molecules that reside inside the cell (whether in the cytoplasm or in the nucleus).

sensitization: A form of nonassociative learning that results after the repeated administration of a stimulus in an amplified response to that stimulus.

sensory-motor neuronal circuits: Circuits of nerve cells that make up the sensory (bringing information into the brain) and the motor (information from the brain to muscles) neuronal systems.

serotonin: A small molecule that acts in both the gastrointestinal tract and the central nervous system (CNS). In the CNS it acts as a neurotransmitter and is responsible for feelings of well-being when released in large amounts in the brain.

signal transduction: The conversion of one kind of molecular signal into another. In neurochemistry the conversion to an action potential of chemical reactions, such as olfactory molecules reacting with olfactory transmitters, is a good example of signal transduction.

small RNA: Small stretches of RNA now thought to be important in regulating genes. These include microRNAs (miRNAs), endogenous small interfering RNAs (esiRNAs), Piwi-interacting RNAs (piRNAs), and promoter-associated RNAs (PASRs).

somatosensory system: A diverse system comprising the receptors and processing parts of the nervous system that produce sensory responses such as touch, temperature, recognition of body position, and pain. Sensory receptors of the somatosensory system are found in bones and joints, covering the skin, in skeletal muscles, and in internal organs and the cardiovascular system.

speech: The means by which language is (usually) expressed by modern humans. It depends on the ability to modulate the vibrating column of air in the upper vocal tract to produce a wide range of sounds that correspond to mental symbols. This ability is generally believed in some way to be a function of the retracted face and lowered larynx typical of adult humans.

split brain phenomenon: The treatment of epilepsy is sometimes accomplished by a process called commissurotomy, a surgical procedure whereby the corpus callosum (the region of the brain connecting the left right hemispheres) is split down the middle. This surgical procedure is effective at limiting the connectivity of the two sides of the brain and at relieving the epilepsy; in the process, however, it also disconnects the left and right hemispheres, giving rise to what are called split brain phenomena.

stable-isotope analysis: A chemical approach to determining the diets of early hominids using stable forms of oxygen and nitrogen often found preserved in their fossilized hard tissues.

stone "core" (or "nucleus"): The lump of stone from which flakes are stuck using a "hammer" of stone (or bone or some similar "soft" substance) to produce a tool with a cutting edge.

synapse: The gap between two neural cells where one neuron sends messages to another. Real neurons can have *thousands* of synapses, allowing them to communicate with thousands of different neurons at once. And each neuron can receive many different *kinds* of chemical messages simultaneously via the synapse. The neural cell that passes on the signal is called the presynaptic neuron, and the cell that receives the impulse is called the postsynaptic neuron.

telencephalon: Also known as the pallium. Along with the diencephalon, this region of the vertebrate brain makes up the forebrain. It is the anteriormost region of the brain and is composed of three major regions known as the cerebrum (the cerebral cortex and prefrontal cortex), the basal ganglia, and the limbic system.

Tetrapoda: Vertebrate animals with four limbs, including mammals, birds, amphibians, all reptiles (including limbless ones such as snakes), and turtles. The marine ancestors of these vertebrates also technically belong to this group.

thalamus: A bilaterally symmetrical paired inner structure of the brain that lies between the cerebral cortex and the midbrain. This region of the brain is considered by some to be part of the limbic system and is essential for relaying sensations, spatial cognition, and motor signals to the cerebral cortex. This region of the brain also regulates alertness and sleep.

transcranial magnetic stimulation (TMS): A technique used to stimulate specific regions of the brain. By rapidly pulsing a magnetic field *outside* the head,

TMS can create electric pulses *inside* the brain. These electric pulses cause neural activity only in targeted areas.

transient receptor potential ion channel (TRP): A category of ion channels located in the membranes of many different cell types. TRPs mediate a wide variety of stimuli and sensations.

transphyletic rescue: The rescue of a wild-type phenotype in an organism that lacks a functional gene for the phenotype by the transfer of a gene from another organism belonging to a different group. For example, the "eyeless" mutation in *Drosophila* can be rescued by transferring the Pax6 gene from mice to the genome of the fly.

Upper Paleolithic: Industry associated with the Cro-Magnons, the earliest modern humans to occupy Europe and the makers of the first "art."

umwelt: The environmental factors that affect an organism's behavior. This term has recently been applied to humans and their practice of naming things by journalist Carol Yoon.

very early hominids: Ancient human relatives of the period between about seven and four million years ago. Currently they include uniquely African species allocated to the genera *Sahelanthropus*, *Orrorin*, and *Ardipithecus*.

Wernicke's area: A patch of the cortex low in the parietal lobe of the brain that is associated with language comprehension.

LITERATURE CITED AND FURTHER READING

Below we list some of the major works consulted in the course of writing this book, including all of those quoted from or cited. For convenience they are grouped by chapter.

Chapter 1. The Nature of Science: Our Brains at Work

Boake, C. R., D. K. Price, and D. K. Andreadis. 1998. "Inheritance of Behavioural Differences between Two Interfertile, Sympatric Species, *Drosophila silvestris* and *D. heteroneura*." *Heredity* 80:642–650.

Eldredge, N., and S. J. Gould. 1972. "Punctuated Equilibria: An Alternative to Phyletic Gradualism." In T. J. M. Schopf, ed., *Models in Paleobiology*, 82–115. San Francisco: Freeman, Cooper.

Gould, S. J., and R. C. Lewontin. 1979. "The Spandrels of San Marco and the Panglossian Paradigm: A Critique of the Adaptationist Programme." *Proceedings of the Royal Society, London*, Ser. B, 205:581–598.

Marcus, G. 2008. *Kluge: The Haphazard Evolution of the Human Mind*. New York: Houghton Mifflin.

Medawar, P. 1969. *Induction and Intuition in Scientific Thought*. London: Methuen.

Popper, K. 1965. *Conjectures and Refutations: The Growth of Scientific Knowledge*. New York: Basic Books.

Yoon, C. K. 2009. *Naming Nature: The Clash between Instinct and Science*. New York: W. W. Norton.

Chapter 2. The Nitty-Gritty of the Nervous System

Arendt, D. 2008. "The Evolution of Cell Types in Animals: Emerging Principles from Molecular Studies." *Nature Reviews: Genetics* 9:868–82.

Bray, D. 2009. *Wetware*. New Haven: Yale University Press.

Chu, N. S. 2006. "Centennial of the Nobel Prize for Golgi and Cajal—Founding of Modern Neuroscience and Irony of Discovery." *Acta Neurologica Taiwanica* 15:217–22.

Crick, F. H. C. 1995. *The Astonishing Hypothesis: The Scientific Search for the Soul*. New York: Scribner.

Crick, F. H., and L. E. Orgel. 1973. "Directed Panspermia." *Icarus* 19:341–348.

Gilbert, W. 1986. "The RNA World." *Nature* 319:618.

Gomperts, B. D., I. Kramer, and M. Tatham. 2002. *Signal Transduction*. Santa Barbara, CA: Academic Press.

Hellman, H. 2001. *Great Feuds in Medicine: Ten of the Liveliest Disputes Ever*. New York: Wiley.

Seung, H. S. 2009. "Reading the Book of Memory: Sparse Sampling versus Dense Mapping of Connectomes." *Neuron* 62:17–29.

Chapter 3. Hanging Our Brains on the Tree of Life

Baluška, F., and S. Mancuso. 2009. "Deep Evolutionary Origins of Neurobiology: Turning the Essence of 'Neural' Upside-Down." *Communicative and Integrative Biology* 2:60–65.

Bear, M. F., B. W. Connors, and M. A. Paradiso. 2007. *Neuroscience: Exploring the Brain*. Philadelphia: Lippincott, Williams and Wilkins.

Bruce, L. L., and T. J. Neary. 1995. "The Limbic System of Tetrapods: A Comparative Analysis of Cortical and Amygdalar Populations." *Brain Behavior and Evolution* 46:224–34.

Chiang, A.-S., et al. 2010. "Three-Dimensional Reconstruction of Brain-Wide Wiring Networks in *Drosophila* at Single-Cell Resolution." *Current Biology*, doi:10.1016/j.cub.2010.11.056.

Chiu, J., et al. 1999. "Molecular Evolution of Glutamate Receptors: A Primitive Signaling Mechanism That Existed before Plants and Animals Diverged." *Molecular Biology and Evolution* 16:826–838.

Darwin, C. 1880. *The Power of Movement in Plants*. London: John Murray.

Davies, E. 2006. "Electrical Signals in Plants: Facts and Hypotheses." In A. G. Volkov, ed., *Plant Electrophysiology: Theory and Methods*, 407–422. New York: Springer.

De Robertis, E. M., and Y. Sasai. 1996. "A Common Plan for Dorsoventral Patterning in Bilateria." *Nature* 380:37–40.

Dunn, C. W., et al. 2008. "Broad Phylogenomic Sampling Improves Resolution of the Animal Tree of Life." *Nature* 452:745–749.

Eitel, M., and B. Schierwater. 2010. "The Phylogeography of the Placozoa Suggests a Taxon-Rich Phylum in Tropical and Subtropical Waters." *Molecular Ecology* 19:2315–2327.

Elliott, G. R. D., and S. P. Leys. 2007. "Coordinated Contractions Effectively Expel Water from the Aquiferous System of a Freshwater Sponge." *Journal of Experimental Biology* 210:3736–3748.

Emes, R. D., et al. 2008. "Evolutionary Expansion and Anatomical Specialization of Synapse Proteome Complexity." *Nature Neuroscience* 11:799–806.

Fitch, W. M. 1970. "Distinguishing Homologous from Analogous Proteins." *Systematic Biology* 19:99–113.

Heimberg, A. M., et al. 2010. "MicroRNAs Reveal the Interrelationships of Hagfish, Lampreys, and Gnathostomes and the Nature of the Ancestral Vertebrate." *Proceedings of the National Academy of Sciences USA* 107:19379–19383.

Holland, N. 2004. "Early Central Nervous System Evolution: An Era of Skin Brains?" *Nature Reviews: Neuroscience* 4:617–627.

Lau, S., G. Jürgens, and I. De Smet. 2008. "The Evolving Complexity of the Auxin Pathway." *Plant Cell* 20:1738–1746.

Meech, R. 2008. "Non-Neural Reflexes: Sponges and the Origins of Behaviour." *Current Biology* 18:R70–R72.

Meyers, P. Z. 2010. "Reinventing the Worm." ScienceBlogs.com/Pharyngula.

Miller, M. B., and B. L. Bassler. 2001. "Quorum Sensing in Bacteria." *Annual Review of Microbiology* 55:165–199.

Nieuwenhuys, R. 2002. "Deuterostome Brains: Synopsis and Commentary." *Brain Research Bulletin* 57:257–270.

Trewavas, A. J. 2007. "Response to Alpi et al.: Plant Neurobiology—All Metaphors Have Value." *Trends in Plant Science* 12:231–233.

Wade, N., 2010. "Decoding the Human Brain, with Help from a Fly." *New York Times*, December 13.

Waters, C. M., and B. L. Bassler. 2005. "Quorum Sensing: Cell to Cell Communication in Bacteria." *Annual Review of Cell and Developmental Biology* 21:319–346.

Westfall, J. A. 1996. "Ultrastructure of Synapses in the First-Evolved Nervous Systems." *Journal of Neurocytology* 25:735–746.

Chapter 4. Making Sense of Senses

Boettiger, A., B. Ermentrout, and G. Oster. 2009. "The Neural Origins of Shell Structure and Pattern in Aquatic Mollusks." *Proceedings of the National Academy of Sciences USA* 106:6837–6843.

Bourne, A. G. 1887. "Sense of Taste or Smell in Leeches." *Nature* 36:125–126.

Damann, N., T. Voets, and B. Nilius. 2008. "TRPs in Our Senses." *Current Biology* 18: R880–R889.

de Bruyne, M., K. Foster, and J. R. Carlson. 2001. "Odor Coding in the *Drosophila* Antenna." *Neuron* 30:537–552.

Gregory, R. L. 2005. *Eye and Brain: The Psychology of Seeing*. London: Weidenfeld and Nicolson.

Keller, A., and L. B. Vosshall. 2004. "A Psychophysical Test of the Vibration Theory of Olfaction." *Nature Neuroscience* 4:337–338.

Lee, Y., et al. 2005. "Pyrexia Is a New Thermal Transient Receptor Potential Channel Endowing Tolerance to High Temperatures in *Drosophila melanogaster*." *Nature Genetics* 37:305–310.

Meshberger, F. L. 1990. "An Interpretation of Michelangelo's 'Creation of Adam' Based on Neuroanatomy." *Journal of the American Medical Association* 264:1837–1841.

Sherkheli, M. A. 2007. "Selective TRPM8 Agonists: A Novel Group of Neurophathic Analgesics." *FEBS Journal* 274 (s1):232.

Shichida, Y., and T. Matsuyama. 2009. "Evolution of Opsins and Phototransduction." *Philosophical Transactions of the Royal Society, London, Ser. B*, 364:2881–2895.

Suk, I., and R. J. Tamarg. 2010. "Concealed Neuroanatomy in Michelangelo's 'Separation of Light from Darkness' in the Sistine Chapel." *Neurosurgery* 66:851–861.

Turin, L. 1996. "A Spectroscopic Mechanism for Primary Olfactory Reception." *Chemical Senses* 21:773–791.

Viana, F., E. la Peña, and C. Belmonte. 2002. "Specificity of Cold Thermotransduction Is Determined by Differential Ionic Channel Expression." *Nature Neuroscience* 5:254–260.

Wakefield, M. J., et al. 2008. "Cone Visual Pigments of Monotremes: Filling the Phylogenetic Gap." *Visual Neuroscience* 25:257–64.

Xiang, Y., et al. 2010. "Light-Avoidance-Mediating Photoreceptors Tile the *Drosophila* Larval Body Wall." *Nature* 468:921–926.

Chapter 5. Processing Information

Bear, M., B. Connors, and W. Paradiso. 1996. *Neuroscience: Exploring the Brain*. Chicago: R. R. Donnelly and Sons.

Emery, N., N. Clayton, and C. Frith. 2007. *From Brain to Culture*. New York: Oxford University Press.

MacLean, P. D. 1990. *The Triune Brain in Evolution: Role in Paleocerebral Functions*. New York: Plenum Press.

Pradel, A., et al. 2009. "Skull and Brain of a 300-Million-Year-Old Chimaeroid Fish Revealed by Synchrotron Holotomography." *Proceedings of the National Academy of Sciences USA* 106:5224–5228.

Putnam, F. W. 1872. "The Blind Cave Fishes of the Mammoth Cave and Their Allies." *Nature* 143:246–249.

Ramsey, E. 1901. "The Optic Lobes and Optic Tracts of *Amblyopsis spelaeus* DeKay." *Journal of Comparative Neurology* 11:40–47.

Reiner, A. 2009. "You Cannot Have a Vertebrate Brain without Basal Ganglia." In H. J. Groenewegen et al., eds., *The Basal Ganglia IX*, 3–24. New York: Springer.

Rose, J. D. "Do Fish Feel Pain?" http://cotrout.org/do_fish_feel_pain.htm.

Smith, J. A. 1991. "A Question of Pain in Invertebrates." *ILAR Journal* 33:1–2.

Strausfeld, N. J. 1998. "The Insect Mushroom Body: A Uniquely Identifiable Neuropil." In J. L. Leonard, ed., *Identified Neurons in Model Systems*. Cambridge, MA: Harvard University Press.

Tomer, R., et al. 2010. "Profiling by Image Registration Reveals Common Origin of Annelid Mushroom Bodies and Vertebrate Pallium." *Cell* 142:800–809.

Chapter 6. Emotions and Memory

Damasio, A. 2003. *Looking for Spinoza*. New York: Harcourt.

Darwin, C. 1872. *The Expression of the Emotions in Man and Animals*. London: John Murray.

Gobet, F., A. de Voogt, and J. Retschitzki. 2004. *Moves in Mind: The Psychology of Board Games*. London: Psychology Press.

Goelet, P., V. F. Castellucci, S. Schacher, and E. R. Kandel. 1984. "The Long and the Short of Long-Term Memory—a Molecular Framework." *Nature* 322:419–422.

Henke, K. 2010. "A Model for Memory Systems Based on Processing Modes Rather than Consciousness." *Nature Reviews: Neuroscience* 11:523–532.

Kepecs, A. "How Addictive Drugs Hijack Your Dopamine System." http://bigthink.com/adamkepecs#!video/.

Müller, N. G., and R. T. Knight. 2006. "The Functional Neuroanatomy of Working Memory: Contributions of Human Brain Lesion Studies." *Neuroscience* 139:51–58.

Müller, N. G., L. Machado, and R. T. Knight. 2002. "Contribution of Subregions of the Prefrontal Cortex to Working Memory: Evidence from Brain Lesions in Humans." *Journal of Cognitive Neuroscience* 14:673–86.

Panskepp, J., 1998. *Affective Neuroscience: The Foundations of Human and Animal Emotions*. Oxford: Oxford University Press.

Squire, L. R., and E. Kandel. 2004. *Memory: From Mind to Molecules*. 2nd ed. Greenwood Village, CO: Roberts.

Chapter 7. Brain EvoDevo

Dasen, J. S., and T. M. Jessell. 2009. "Hox Networks and the Origins of Motor Neuron Diversity." *Current Topics in Developmental Biology* 88:1–32.

Finarelli, J. A., and J. J. Flynn. 2009. "Brain-Size Evolution and Sociality in Carnivora." *Proceedings of the National Academy of Sciences USA* 106:9345–9349.

Gould, S. J. 1977. *Ontogeny and Phylogeny.* Cambridge, MA: Harvard University Press.

Haeckel, E. 1874. *Anthropogenie oder Entwickelungsgeschichte des Menschen: Gemeinverständliche wissenschaftliche Vorträge über die Grundzüge der menschlichen Keimes- und Stammes-Geschichte.* Leipzig: Engelmann.

Khaitovich, P., W. Enard, M. Lachmann, and S. Pääbo. 2006. "Evolution of Primate Gene Expression." *Nature Reviews: Genetics* 77:693–702.

King, M.-C., and A. C. Wilson. 1975. "Evolution at Two Levels in Humans and Chimpanzees." *Science* 188:107–116.

Levine, J. S., and K. R. Miller. 1994. *Biology: Discovering Life.* Boston: D. C. Heath.

Mattick, J. S., and M. F. Mehler. 2008. "RNA Editing, DNA Recoding and the Evolution of Human Cognition." *Trends in Neuroscience* 31:227–233.

Montgomery, S., et al. 2010. "Adaptive Evolution of Four Microcephaly Genes and the Evolution of Brain Size in Anthropoid Primates." *Molecular Biology and Evolution* 28:625–639.

Preuss, T. M., M. Caceres, M. C. Oldham, and D. H. Geschwind. 2004. "Human Brain Evolution: Insights from Microarrays." *Nature Reviews: Genetics* 5:850–860.

Sol, D., et al. 2010. "Evolutionary Divergence in Brain Size between Migratory and Resident Birds." *PLoS One* 5(3):e9617.

Weisbecker, V., and A. Goswami. 2010. "Brain Size, Life History, and Metabolism at the Marsupial/Placental Dichotomy." *Proceedings of the National Academy of Sciences USA* 107:16216–16221.

Chapter 8. Words and Music by . . .

Bolhuis, J. J., K. Okanoya, and C. Scharff. 2010. "Twitter Evolution: Converging Mechanisms in Birdsong and Human Speech." *Nature Reviews: Neuroscience* 11:747–759.

Carleton, K. L., T. C. Spady, and T. Kocher. 2005. "Visual Communication in East African Cichlid Fishes: Diversity in a Phylogenetic Context." In B. G. Kapoor et al., eds., *Fish Communication.* Enfield, NH: Science Publisher.

Corballis, P. M., M. G. Funnell, and M. S. Gazzaniga. 2002. "Hemispheric Asymmetries for Simple Visual Judgments in the Split Brain." *Neuropsychologia* 40:401–410.

Ethnologue. http://www.ethnologue.com/.

Fedorenko, E., and N. Kanwisher. 2009. "Neuroimaging of Language: Why Hasn't a Clearer Picture Emerged?" *Language and Linguistics Compass* 3/4:839–865.

Greenhill, S. J., Q. D. Atkinson, A. Meade, and R. D. Gray. 2010. "The Shape and Tempo of Language Evolution." *Proceedings of the Royal Society, London, Ser. B.,* 277:2443–2450.

Grodzinsky, Y. 2010. "The Picture of the Linguistic Brain: How Sharp Can It Be? Reply to Fedorenko and Kanwisher." *Language and Linguistics Compass* 4:605–22.

Hauser, M. D., N. Chomsky, and W. T. Fitch. 2002. "The Faculty of Language: What Is It, Who Has It, and How Did It Evolve?" *Science* 298:1569–1579.

Levitin, D. 2006. *This Is Your Brain on Music: The Science of a Human Obsession*. New York: Dutton/Penguin.

Mathger, L. M., N. Shashar, and R. T. Hanlon. 2009. "Do Cephalopods Communicate Using Polarized Light Reflections from Their Skin?" *Journal of Experimental Biology* 212:2133–2140.

Price, C. J. 2010. "The Anatomy of Language: A Review of 100 fMRI Studies Published in 2009." *Annals of the New York Academy of Sciences* 1191:62–88.

Rizzolatti, G., and L. Craighero. 2004. "The Mirror-Neuron System." *Annual Review of Neuroscience* 27:169–192.

Vargha-Khadem, F., G. D. Gadian, A. Copp, and M. Mishkin. 2005. "FOXP2 and the Neuroanatomy of Speech and Language." *Nature Reviews: Neuroscience* 6:131–138.

Chapter 9. Behaviors and Beliefs

Dawkins, R. 2006. *The God Delusion*. Boston: Houghton Mifflin.

Deaner, R. O., A. V. Khera, and M. L. Platt. 2005. "Monkeys Pay per View: Adaptive Valuation of Social Images by Rhesus Macaques." *Current Biology* 15:543–548.

Forrest, D. V. 2008. "Alien Abduction: A Medical Hypothesis." *Journal of the American Academy of Psychoanalysis and Dynamic Psychiatry* 36:431–442.

Glimcher, P., M. Dorris, and H. Bayer. 2005. "Psychological Utility Theory and the Neuroeconomics of Choice." Working paper, New York University.

Lavallee, C. F., M. A. Persinger, and A. Loret. 2010. "A Study of Mental Time Travel: Similar and Distinct Electrophysiological Correlates of Re-experiencing Past Events and Pre-experiencing Future Events." *Consciousness and Cognition* 19:1037–1044.

Sally, D., and E. Hill. 2006. "The Development of Interpersonal Strategy: Autism, Theory-of-Mind, Cooperation and Fairness." *Journal of Economic Psychology* 27:73–97.

Sanfey, A. G., et al. 2003. "The Neural Basis of Economic Decision-Making in the Ultimatum Game." *Science* 300:1755–1758.

Solomon, A. 2001. *The Noonday Demon: An Atlas of Depression*. New York: Scribner.

Chapter 10. The Human Brain and Cognitive Evolution

Aiello, L., and P. Wheeler. 1995. "The Expensive-Tissue Hypothesis: The Brain and the Digestive System in Human and Primate Evolution." *Current Anthropology* 36:199–221.

Allman, J. 1999. *Evolving Brains*. New York: Freeman.

Amunts, K., et al. 2010. "Broca's Region: Novel Organization Principles and Multiple Receptor Mapping." *PLoS Biology* 8:e1000489.

Arsuaga, J.-L., J. M. Bermudez de Castro, and E. Carbonell, eds. 1997. "Special Issue: The Sima de los Huesos Hominid Site." *Journal of Human Evolution* 33:105–421.

Bar-Yosef, O., and J.-G. Bordes. 2010. "Who Were the Makers of the Châtelperronian Culture?" *Journal of Human Evolution* 59:586–593.

Bocherens, H., et al. 2005. "Isotopic Evidence for Diet and Subsistence Pattern of the Saint-Césaire I Neanderthal: Review and Use of a Multi-Source Mixing Model." *Journal of Human Evolution* 49:71–87.

Briggs, A. W., et al. 2009. "Targeted Retrieval and Analysis of Five Neanderthal mtDNA Genomes." *Science* 325:318–321.

Brown, K. S., et al. 2009. "Fire as an Engineering Tool of Early Modern Humans." *Science* 325:859–862.

Carbonell, E., et al. 2010. "Cultural Cannibalism as a Paleoeconomic system in the European Lower Pleistocene." *Current Anthropology* 51:539–549.

Cohen, J. 2010. *Almost Chimpanzee: Searching for What Makes Us Human in Rainforests, Labs, Sanctuaries and Zoos*. New York: Times Books.

Coolidge, F. L., and T. Wynn. 2009. *The Rise of Homo sapiens: The Evolution of Modern Thinking*. New York: Wiley-Blackwell.

Coqueugniot, H., J.-J. Hublin, F. Veillon, F. Houët, and T. Jacob. 2004. "Early Brain Growth in Homo erectus and Implications for Cognitive Ability." *Nature* 431:299–302.

Dean, C., et al. 2001. "Growth Processes in Teeth Distinguish Modern Humans from Homo erectus and Earlier Hominins." *Nature* 414:628–631.

Dean, M. C., and B. H. Smith. 2009. "Growth and Development of the Nariokotome Youth, KNM-WT 15000." In F. E. Grine et al., eds., *The First Humans: Origin and Early Evolution of the Genus Homo*, 101–120. Heidelberg: Springer.

Geschwind, N. 1964. "The Development of the Brain and the Evolution of Language." *Monograph Series on Languages and Linguistics* 17:155–169.

Gibbons, A. 2006. *The First Human: The Race to Discover Our Earliest Ancestors*. New York: Doubleday.

Harcourt-Smith, W. E. H. 2007. "The Origins of Bipedal Locomotion." In W. Henke and I. Tattersall, eds., *Handbook of Paleoanthropology*, 3:1483–1518. Heidelberg: Springer.

Hart, D., and R. W. Sussman. 2009. *Man the Hunted: Primates, Predators, and Human Evolution.* Expanded ed. Boulder, CO: Westview Press.

Henshilwood, C. S., et al. 2002. "Emergence of Modern Human Behavior: Middle Stone Age Engravings from South Africa." *Science* 295:1278–1280.

———. 2004. "Middle Stone Age Shell Beads from South Africa." *Science* 304:404.

Holloway, R. L., D. C. Broadfield, and M. S. Yuan. 2004. *The Human Fossil Record*, vol. 3: *Brain Endocasts—the Paleoneurological Evidence.* New York: Wiley-Liss.

Klein, R. 2009. *The Human Career: Human Biological and Cultural Origins.* 3rd ed. Chicago: University of Chicago Press.

Lalueza-Fox, C., et al. 2010. "Genetic Evidence for Patrilocal Mating Behavior among Neandertal groups." *Proceedings of the National Academy of Sciences USA*, doi/10.1073/pnas.1011533108.

Lieberman, P. 2007. "The Evolution of Human Speech: Its Anatomical and Neural Bases." *Current Anthropology* 48:39–66.

Lordkipanidze, D., et al. 2005. "The Earliest Toothless Hominin Skull." *Nature* 434:717–718.

———. 2007. "Postcranial Evidence of Early *Homo* from Dmanisi, Georgia." *Nature* 449:305–310.

Plummer, T. 2004. "Flaked Stones and Old Bones: Biological and Cultural Evolution at the Dawn of Technology." *Yearbook of Physical Anthropology* 47:118–164.

Povinelli, D. J. 2004. "Behind the Ape's Appearance: Escaping Anthropocentrism in the Study of Other Minds." *Daedalus* 133 (1):29–41.

Pruetz, J. D., and P. Bertolani. "Savanna Chimpanzees, *Pan troglodytes verus*, Hunt with Tools." *Current Biology* 17:412–417.

Schick, K., et al. 1999. "Continuing Investigations into the Stone Tool-Making and Tool-Using Capabilities of a Bonobo (*Pan paniscus*)." *Journal of Archaeological Science* 26:821–832.

Schwartz, J. H., and I. Tattersall. 2002. *The Human Fossil Record*, vol. 1: *Terminology and Craniodental Morphology of Genus* Homo *(Europe).* New York: Wiley-Liss.

Sponheimer, M., and J. Lee-Thorp. 2007. "Hominin Paleodiets: The Contribution of Stable Isotopes." In W. Henke and I. Tattersall, eds., *Handbook of Paleoanthropology*, 555–585. Heidelberg: Springer.

Tattersall, I. 2008. "An Evolutionary Framework for the Acquisition of Symbolic Cognition by *Homo sapiens*." *Comparative Cognition and Behavior Reviews* 3:99–114.

———. 2009. *The Fossil Trail: How We Know What We Think We Know about Human Evolution.* 2nd ed. New York: Oxford University Press.

———. 2011. "Origin of the Human Sense of Self." In W. van Huyssteen and E. B. Wiebe, eds., *In Search of Self.* Chicago: Wm. B. Eerdmans.

Thieme, H. 1997. "Lower Palaeolithic Hunting Spears from Germany." *Nature* 385:807–810.

Walker, A. C., and R. E. F. Leakey. 1993. *The Nariokotome* Homo erectus *Skeleton.* Cambridge, MA: Harvard University Press.

Wood, B., and M. Collard. "The Human Genus." *Science* 284:65–71.

Wrangham, R. 2009. *Catching Fire: How Cooking Made Us Human.* New York: Basic Books.

INDEX

Page numbers in *italics* refer to illustrations.

chicken, embryonic development in, 198

chimpanzees, 71, 103, 113, 114, 197, 226,
227, 264, 274, 275; behavior, 234–35,
260–62; brain, 202–5, 259–62; cognition, 259–62; communication, 234–35,
260–61; -human differences, 197, 202–5,
214–15, 261–62

chloride, 74

chlorine, 35, 43

chloroplasts, 37, 143

choanocytes, 76

choanoflagellates, 78, 80, 97

Chomsky, Noam, 233

chordates, 90–93, 145, 147; brains, 90–93,
147–48

Christianity, 18, 19, 248

chromophores, 231–32

chromosomes, 16, 108, 213, 214, 224–25,
296; translocation, 224

chunking, 180

cichlids, 112, 231

cilia, 66, 77, 124, 125, 125

ciliary opsins, 108, 109, 110, 112, 114

ciliates, 66

cingulate gyrus, 157, 256

circadian rhythm, 108

clams, 85, 128

Class, 13

classification, 11–14, 61, 85

Clayton, Nicola, 159

cleaver, 279

climate, 264, 286; fluctuations, 264,
285–86, 289, 295–97, 308; Ice Age,
285–86, 289, 290, 307, 308

clinic-anatomical approach, 182, 186–90,
219

clothing, 289

Cnidaria, 78–80, 82, 95, 96, 97, 97, 104,
127–28, 135; nerve nets and sensing,
144–46

cnidocytes, 78

cochlea, 115, 116, 117, 136, 137

coelom, 81, 82

coelurosaurs, 158–59

cognition, 217, 239; ape, 234–35, 259–62;
australopiths, 267–76; brain size and,
210–15, 259, 270–71, 271; decisions,
behaviors, and beliefs, 239–53; early
hominid, 259–67, 268, 269–71, 271,
272–82; emergence of modern
behavior, 292–300, 305–6; *Homo
ergaster*, 276–80; *Homo heidelbergensis*,
282–86; human brain and cognitive
evolution, 255–300; language and
music, 217–38; Neanderthals and
Cro-Magnons, 286–92, 297; starting
point, 259–62; stone tool-making
and, 272–80; symbolic, 260–61, 286,
290–92, 292, 294–97, 299, 302,
305, 307

color, 143–44, 241; change, 231–32;
vision, 108, 109, 110, 111, 112, 114, 149

colossal squid, 57

Columbia University, 182, 184, 200, 249

comb jellies, 135

communication, 217–38, 260; animal,
231–35; cell-to-cell, 56–60, 62, 63,
68–75, 79, 94–97, 117, 177; language,
217–35, 297–300; leap to language,
297–300; left-right brain, 219–21;
music, 235–38; visual means of,
231–32

compound eye, 105, 106, 106

conditioning, 186

connectome, 58, 59, 60

Connors, Barry W., 158

consciousness, 27, 145

contact chemoreception, 118

convergence, 7–8, 9, 14, 17, 75, 99, 178,
229, 240

Coolidge, Frederick, 298, 299

corals, 78

mammals, 93, 178; avian brain and,
158–59, 160; brains, 89, 146, 156–59,
160, 174, 257, 258; emotions, 173–79;
grazing, 266, 275; motor movement,
156; placental, 211; sense systems in,
104, 111, 112, 133–34; *see also specific
mammals*
Mancuso, Stefano, 74
Marcus, Gary, 4, 304
marsupials, 211
Matsuyama, Take, 108
Mattick, John, 210
maxillary palps, 122–23
Mayr, Ernst, 105
mazes, 67, 68, 68
meat eating, 212, 275–76, 284
mechanoreceptors, 128–30, 130, 131–34
Medawar, Peter, 7
medial temporal lobe, 188
medulla, 149, 160
medusae, 79
Meech, Robert, 76
Mehler, Mark, 210
Meissner's corpuscles, 128–29, 130
melanopsin, 108
membranes, 39–45, 49, 66, 131, 132,
257; receptors, 49–53, 53, 54–55, 55;
scaffolding, 95
memory, 5, 57, 142, 153, 171, 179–96,
210, 225; brain injuries and, 186–90;
declarative, 181, 189, 190, 193; episodic,
181; long-term, 179–80, 188–95, 256;
LTP and, 191–95; nervous system and,
184–86; nondeclarative, 181, 182, 183,
184, 188, 193; priming, 182; procedural, 181–82, 188–89; sea slug, 182–86;
semantic, 181; sensory, 179–80;
short-term, 179–80, 188, 190; synapses,
190–95; working, 189–90, 299
menthol, 133, 134
Merkel's disks, 128–29, 130

mesoderm, 128
mesoglea, 128
Mesozoa, 86
metabolic regulation, 178
metabotropic receptor, 52, 71, 72
Metazoa, 178
meteors, 37
Meyers, P. Z., 84–85
mice, 57, 58, 87, 95; brain, 58, 59, 60,
193–94; memory experiments, 193–94
Michelangelo, 102–3
microarray, 203, 208, 214
microcephaly, 212–15, 219, 223
microtubes, 49
midbrain, 84, 149, 161, 199, 200
Middle Stone Age, 293–95
Miller, Ken, 198, 199
Miller, Melissa, 66
Milner, Brenda, 188
mimosas, 69
miRNAs, 47, 206–7
miR-124, 207–8
mirror neurons, 237–38
mitochondria, 37, 69, 296
mitosis, 88
mitral cells, 125, 125
model organism approach, 182
Molaison, Henry Gustav, 188–89
molecular biology, 25
molecular fingerprint, 46–47
molecular systematics, 11
molecules, 36, 38–41, 49, 63, 105;
odorant, 121–27; signal transduction,
48–55; taste, 119–21
Molière, *Candide*, 21
mollusks, 85, 106, 108, 128, 145; ear,
116; taste, 118–19
molting, 85
monkeys, 113, 114, 232, 237, 259, 274;
brains, 244–45; decision-making,
244–45; mirror neurons, 237

monoamines, 165

Monosiga, 97

Montgomery, Stephen, 214

morphogen, 202

Morris water maze, 193

mosses, 8

motor movement, 156

motor neurons, 57

motor skills, 227, 274, 278, 279, 289

mouth, 119

mucus, 124, 128

Muller, Notger, 189, 190

multicellularity, 63, 66, 78, 80; evolution
of, 66–69

mu opioid, 169

muscles, 80, 129, 131

mushroom body, 146–47

music, 217, 235–38, 291

mutation, 24–25, 132, 207, 213, 227;
fruit fly, 132

myelin, 424

myoepithelial cells, 47

myriapods, 147

myxomycetes, 66, 67

Myxozoa, 86

nasal epithelium, 124, 125, 125

nasal passage, human, 123–24, 124

natural selection, 18, 20–26, 39, 253,
265, 303, 305; adaptation and, 20–26

Neanderthals, 202, 226–28, 236–37, 268,
271, 282, 286–87, 288, 289–92, 293,
297, 304; cognition and behavior,
286–92, 297–98; decline of, 289,
291; skeleton, 288

nematodes, 57, 81–84, 85, 103, 104,
108–9, 131, 182, 206; brains, 81, 82,
83; larvae, 109; nervous system, 82,
83, 90; senses, 108–9, 119, 121–23,
134; smell, 121–22, 122, 123; taste, 119

Nemertea, 85

neocortex, 153–55, 182, 212

neoteny, 205

nervous system, 29–60, 63, 69, 82,
89–93, 94–97, 103, 132, 149, 173, 199;
balance and, 137–39; bilaterian, 81–85,
96–98; cell communication, 56–59, 59,
60, 63, 79, 95–97, 117; central, 87, 182,
207; chordate, 91–93; deuterosome,
81–82, 91–92; echinoderm, 90–91;
electrical impulses and, 41–46, 48,
58; embryonic development, 197–201,
201; hearing and, 114–17; hemichordate, 91; ions and, 31–36; memory
and, 184–86; nematode, 82, 83, 90;
peripheral, 163, 182; plants and,
69–75, 127; sea slug, 184–85; signal
transduction and, 48–55; smell and,
121–27; sponges and, 75–78, 76,
95–97; synapses and, 94–97

neural net, 78–80, 95, 97, 128, 144–46;
Cnidarian, and sensing, 144–46

neural tubes, 207; development of,
200–202, 207

neuroeconomics, 239–47

neuroimaging, 189–90, 219, 223, 225,
246, 252

neuromodulators, 167, 168, 169

neuron doctrine, 56

neurons, 43, 45, 51, 57, 60, 88, 89, 105,
184–85, 190, 205, 258; mirror, 237–38;
motor, 57; sensory, 57

neuropsins, 108

neurotransmitters, 44, 51–54, 70–73,
80, 163–71, 191; clean up, 167–69;
systems, 166

neutrons, 30, 32, 33

New World monkeys, 113, 114

9/11 attacks, 181, 196

nitrogen, 35

NMDA, 71, 72; receptors, 191–95

nociception, 130, 131, 153–55